Elizabeth A. Wood

CRYSTALS AND LIGHT

An Introduction to Optical Crystallography

(Second Revised Edition)

Dover Publications, Inc.
New York

To Ida H. Ogilvie and
Dorothy Wyckoff

Preface

This is a book about the behavior of light in crystals. Since it is written for people with no previous training in either the subject of crystals or the subject of light, the first six chapters tell what sort of thing a crystal is and the seventh tells about some of its physical properties. The remaining chapters describe some of the effects produced on light, mostly polarized light, by crystals and so constitute an introduction to optical crystallography.

In this subject, good color illustrations are essential, not only because they save many words of description, but more importantly, because they give the reader that familiarity with the hues and intensities which will enable him to interpret the optical effects produced by crystals between crossed polarizers. Therefore it is a pleasure to acknowledge my indebtedness to Ernst Leitz GMBH, Wetzlar, for underwriting the cost of the color plates and for providing the color photographs of interference figures and of quartz and calcite particles between crossed polarizers. Without this assistance it would have been impossible to include color illustrations in this book. I am also grateful to S. O. Jorgensen of the Bell Telephone Laboratories for the remaining color photo-

---

Published in Canada by General Publishing Company, Ltd., 30 Lesmill Road, Don Mills, Toronto, Ontario. Published in the United Kingdom by Constable and Company, Ltd., 10 Orange Street, London WC2H 7EG.

This Dover edition, first published in 1977, is a revised republication of the work originally published by D. Van Nostrand Company, Inc., in 1964.

International Standard Book Number: 0-486-23431-2
Library of Congress Catalog Card Number: 76-27458
Manufactured in the United States of America
Dover Publications, Inc.
180 Varick Street
New York, N. Y. 10014

graphs. None of the photographs would have come through satisfactorily to the reader without the conscientious attention of the publisher and printer to high-quality rendition of the color. Most of the more intricate line drawings were done by F. M. Thayer of the Bell Telephone Laboratories, to whom I am grateful for their skillful execution.

Several people have read all or part of this book, and it has benefited from their suggestions. First among these are Melba Phillips of the University of Chicago, R. L. Barns, and A. N. Holden of the Bell Telephone Laboratories, who read the entire book and suggested corrections and additions of importance. Others who have contributed in this way are E. U. Condon, I. Fankuchen, V. B. Compton, W. L. Bond, I. D. Payne, and P. Singer. My husband, I. E. Wood, has not only critically read parts of the manuscript, but has also provided that understanding cooperation without which a married author cannot enjoy writing a book.

This book is dedicated to Ida H. Ogilvie and Dorothy Wyckoff. These are the talented teachers from whom it was my good fortune to receive my early education in crystallography.

At the 1962 meeting of the American Physical Society in Baltimore, J. D. H. Donnay, well known among crystallographers, began a discourse on the significance of crystal forms with the statement, "One should not turn up one's nose at facts simply because they can be seen with the naked eye." In these days of pions and muons and strange particles, this is a thought-provoking admonition.

Many of the facts presented in this book can be seen with the naked eye, the reader's eye. May they give him pleasure.

This edition has benefited from suggestions made by several readers, but especially by those of Professors P. P. Ewald and J. D. H. Donnay. Professor Donnay's extremely careful reading of the text has resulted in many improvements in the present edition.

ELIZABETH A. WOOD

Table of Contents

[Plates I–VIII follow page 28.]

1 *Symmetry*

A crystal is a solid composed of atoms arranged in an orderly repetitive array. That is about the shortest definition you will find of a crystal. Some will consider it incomplete, but all will agree that what it says is true.

Consider, for example, the very widely known crystal calcite, calcium carbonate ($CaCO_3$), which, because it is found in nature, is called a mineral. The metamorphic rock, marble, is made up of very small interlocking crystals of calcite, and so is the sedimentary rock, limestone. Most white crushed-stone driveways are made of bits of calcite, and a crystal of calcite was the essential part of an optical ring sight used in World War II. Later we will see how this ring sight worked.

The pattern on the inside of the front cover shows the arrangement of calcium, carbon, and oxygen atoms in calcite magnified about a hundred million times. Each black carbon atom is symmetrically surrounded by three white oxygen atoms. If we were to erect an imaginary axis normal (perpendicular) to the paper through the center of this CO_3 group and rotate the group around it, the oxygen atoms would occupy indistinguishable positions on the paper three times during a complete revolution. Therefore this is "an axis of 3-fold symmetry," or "a 3-fold axis" for short.

Suppose we rotate the whole pattern around this imaginary axis. If this is difficult to imagine, we can place a sheet of tracing paper over the diagram, trace a few of the atoms of the pattern, and then, with a pin at the chosen axis, rotate the tracing paper. We find that the axis is an axis of 3-fold symmetry for the whole pattern. We can choose the axis through any one of the carbon atoms with the same result: they are all "crystallographically equivalent," which means that each one has exactly the same surroundings as any other one. We will return to this pattern later.

Plate II is a photograph of a group of calcite crystals. Such a crystal has grown by the gradual accumulation of atoms, layer upon layer, in an orderly array, the atoms of each layer pulled into the proper arrangement by the attracting forces of those atoms already in place. In the photograph, a crystal in the middle of the group with its edges outlined by ink dots is oriented with its sharp tip toward the observer, its long axis normal to the page. It is down this axis that we are looking in the pattern on the inside of the cover, and we see in the photograph that this is a 3-fold symmetry axis for the outer shape of the crystal. This fact is not surprising because the crystal owes its shape to the orderly array of the atoms which compose it.

As the crystal grows, the supply of material may happen to be greater on one side than another, so that more layers will be added there. The crystal symmetry may thus appear "distorted," but measurement of the angles that the faces make with each other will show that these angles remain the same regardless of the relative sizes of the faces. Examples will be discussed at the beginning of Chapter 4.

This constancy of interfacial angles was one of the first facts about crystals to be recorded. In 1669 Nicolaus Steno, a Danish professor of anatomy, reported that he had measured the angles between corresponding pairs of faces on many different crystals of quartz from different localities and had made the remarkable discovery that they were the same. Robert Hooke, in his *Micrographia* (1665), had attempted to explain the "regular Figures" of crystals by packing together "globular bullets." These, packed as in Fig. 1-1, had the same triangular outline as a face on an alum crystal, he pointed out.

A sense of symmetry is a powerful tool for the study of crystals. It is a simplifying key to the endlessly various arrays of atoms which make up crystalline solids, enabling us to think of them in terms of familiar patterns.

There are only a few symmetry operations. A symmetry operation is an operation such as rotation, performed on an object or pattern, which brings it into coincidence with itself. For example, if the object in Fig. 1-1 is rotated 120° around an axis normal to the paper at its center point, then, as a result of this operation, it occupies a position on the page which is absolutely indistinguishable from the position it occupied originally. This operation is therefore a symmetry operation. Since $3 \times 120° = 360°$, this operation would occur three times during a full rotation, and the axis is therefore a 3-fold symmetry axis.

We can generalize by saying that if an object has an n-fold axis of symmetry, it occupies the same position in space after each angular rotation of $360°/n$ around that axis (e.g., $120° = 360/3$). In Fig. 1-1, we may consider another axis passing through the center, but lying in the plane of the paper. Let it go right between two of the circles (tangent to both) and bisect the third. In order to rotate the figure around this axis we must let it swing into the third dimension out of the 2-dimensional page, but if we rotate it 180° the figure will again be lying in the plane of the paper, occupying a position just like its original one. So this, too, is a symmetry operation of this object. Since $180° = 360°/2$, this is an axis of 2-fold symmetry. How many of these are there in this figure? We could choose one through the center bisecting each of the three circles. There are three such axes, as there must be if one of them is to rotate into another about the 3-fold axis which is normal to the paper at the center of the figure. Fig. 1-2

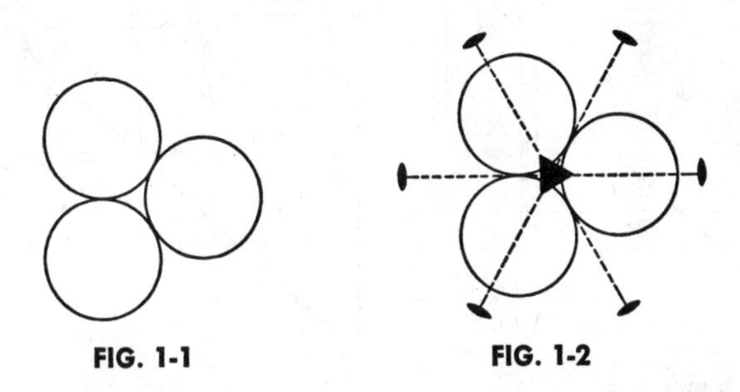

FIG. 1-1 FIG. 1-2

shows these 2-fold axes, tipped with two-cornered symbols to identify them and, coming out of the paper toward us, the 3-fold axis, tipped with a three-cornered symbol. We say the 2-fold axes are consistent with the operation of the 3-fold axis, just as the 3-fold axis is consistent with the operation of the three 2-fold axes: one end of it is just like the other, so that a rotation of 180° around any one of the 2-fold axes will bring it into a position indistinguishable from its initial position.

Since we have already moved into the third dimension, let us explore the symmetry of some familiar three-dimensional objects.

Consider a square four-legged table, like a card table. If we pass an axis vertically through its center like the shaft of a large umbrella over a table in an outdoor restaurant, we find that this is an axis of 4-fold symmetry of the table. Unlike Fig. 1-1, the table has no symmetry axes normal to its 4-fold axis because it has legs going down, but none going up: even disregarding the umbrella, you would have to turn the table 360° around a horizontal axis before it coincided with its original position in space. This can, of course, be called an axis of 1-fold symmetry, but that really means an axis of no symmetry at all.

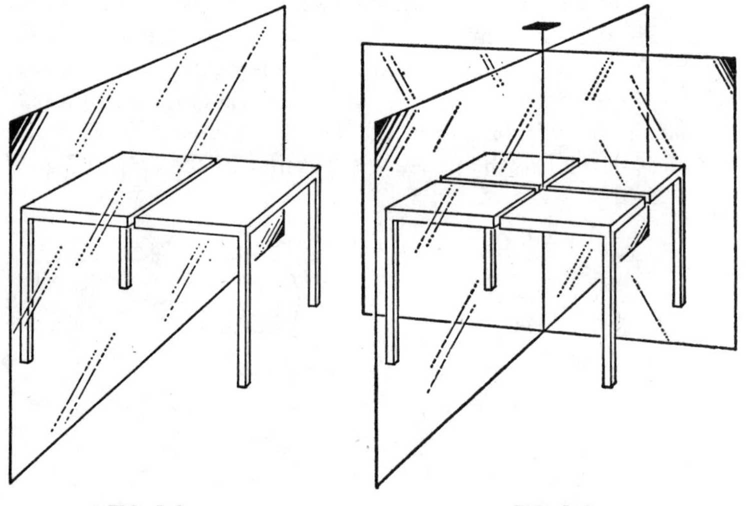

FIG. 1-3 FIG. 1-4

The 4-fold axis is not the only symmetry element of the table. There is another sort of symmetry element present. Suppose we sawed the table in two across the middle, parallel to one side, and then brought the cut edge up against a vertical mirror, as in Fig. 1-3. The table would appear whole again. The mirror image of the right-hand half exactly coincides with the position occupied by the left-hand half, and vice versa. Therefore a vertical mirror plane through the center of the table from the center of one side to the center of the opposite side is a symmetry element of the table. Expressed in terms of three mutually perpendicular axes, x, y and z, with y and z lying in the plane of the mirror, any feature of the table that occurs at a point $+x_1$, $+y_1$, $+z_1$ also occurs at a point $-x_1$, $+y_1$, $+z_1$. To have only one such plane would violate the 4-fold axis, and indeed, without having to saw another table apart, we can see that there would be a second vertical mirror plane of symmetry at right angles to the first, as in Fig. 1-4.

There are also vertical mirror planes (sometimes called symmetry planes) that cut the table diagonally, through opposite corners, so that from the top the complete collection of symmetry elements of the table looks like Fig. 1-5, in which the thin lines represent the mirror planes and the small black square is the four-cornered symbol on the tip of the 4-fold axis.

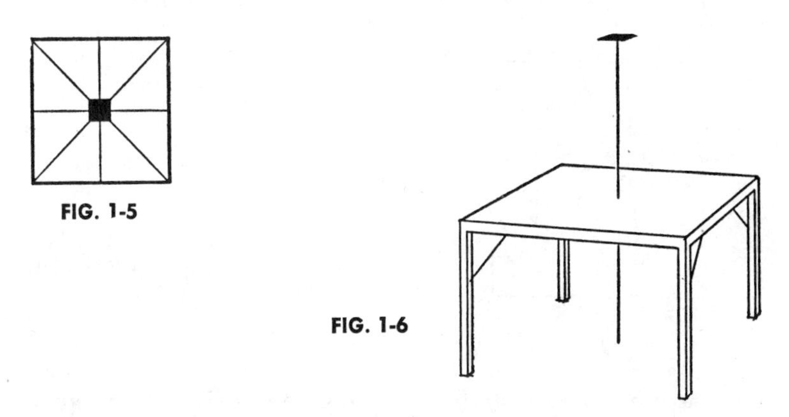

FIG. 1-5

FIG. 1-6

Does the table have a 2-fold axis of symmetry? If you rotate it 180° around the 4-fold axis, its position coincides with the original position, so it has a 2-fold axis which is, so to speak, included in the 4-fold axis. Since there will always be a 2-fold axis coinciding with the 4-fold axis, we mention only the axis of higher symmetry.

Suppose the table has a metal brace to hold each leg, as shown in Fig. 1-6. If we take these into account, the table loses all its planes of symmetry, although it still has a 4-fold axis.

There is a third kind of symmetry element in addition to planes and axes. We can illustrate it with a brick. (See Fig. 1-7.) Most bricks have some sort of markings and irregularities, but let us imagine an ideal brick with every face smooth and un-

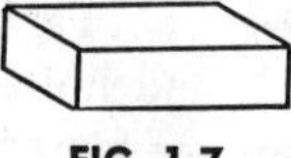

FIG. 1-7

blemished. What symmetry elements does it have? It has mirror planes and axes. Try to determine how many of each and where they are before reading any further. You can use a box that has no square faces as an *ersatz* brick to help you visualize the problem.

Fig. 1-8 shows the symmetry planes and axes of the brick. We

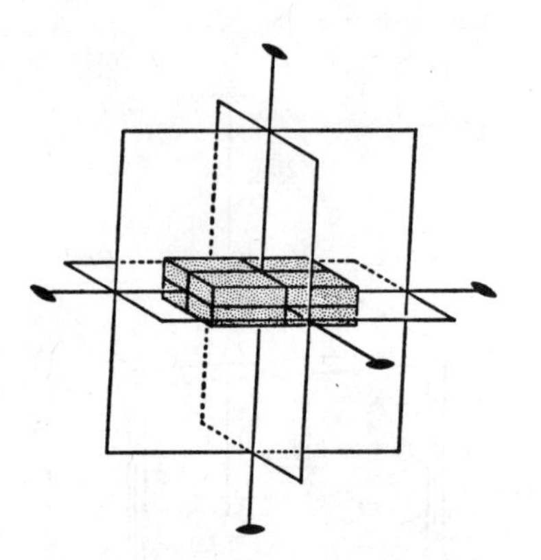

FIG. 1-8

FIG. 1-9

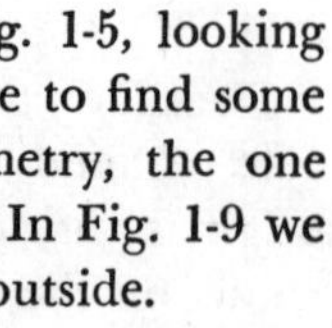

can make a diagram of them like the one in Fig. 1-5, looking down on the brick from the top, but we will have to find some way of indicating the horizontal plane of symmetry, the one normal to the direction in which we are looking. In Fig. 1-9 we have indicated this plane by the circle around the outside.

The brick has an additional symmetry element: a *center* of symmetry. Whatever feature of the brick occurs a given distance away from the exact center in one direction will also occur the same distance away in the opposite direction. If we draw a line from the center of the brick to one corner, then that same line, extended in the opposite direction from the center, will meet another corner at the same distance from the center, as shown in Fig. 1-10.

FIG. 1-10

If we keep the center fixed and move the left end of the line toward ourselves along the edge, the right end will move away from us along the opposite edge, both line segments changing length together, always pivoting on the center of symmetry as in Fig. 1-10. In this way we could proceed along the entire surface of the brick; the two segments of the line passing through the center and ending at the faces, edges, and corners of the brick would always be the same length. This is the test for a center of symmetry in a brick. The "operation" is called *inversion* through a center. Expressed in terms of three mutually perpendicular axes, x, y and z, with their origin at the center, for any feature at x_1, y_1, z_1, there will be an exactly similar feature at $-x_1$, $-y_1$, $-z_1$.

The table did not have a center of symmetry, and neither did the three-circle pattern in Fig. 1-1.

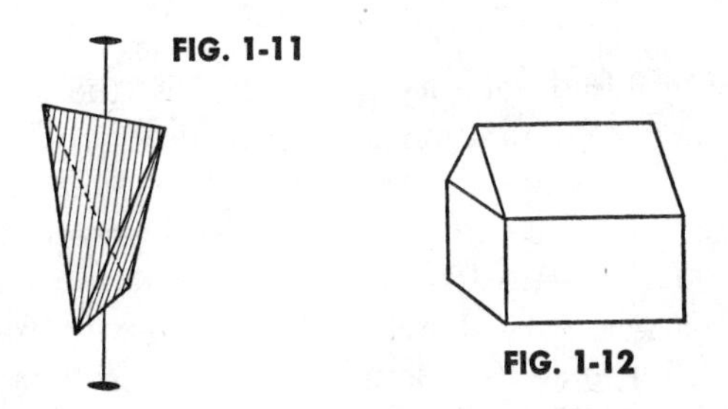

FIG. 1-11

FIG. 1-12

In Fig. 1-11 we see a solid object bounded by four isosceles triangles, called a disphenoid. It has a 2-fold axis of symmetry through the centers of the short edges, and two symmetry planes intersecting in this axis are normal to each other. But so also does the house-like object in Fig. 1-12, which, unlike the sphenoid, does not have four similar surfaces. We need an additional symmetry element to indicate that the disphenoid has more symmetry than the "house." If you rotate the disphenoid 90° around its 2-fold axis and then perform the operation of inversion through the center-point of the object, it will occupy its original position again. This combined rotation and inversion is thus a symmetry operation, and the symmetry element that represents it is called a rotatory inversion axis, or simply an *inversion axis.** In this case it is a 4-fold inversion axis. In diagrams it is convenient to distinguish rotatory inversion axes from rotation axes by using open squares or triangles instead of the solid figures we have been using. Looking down on the disphenoid from the top, we would see the symmetry elements shown in Fig. 1-13, the open square

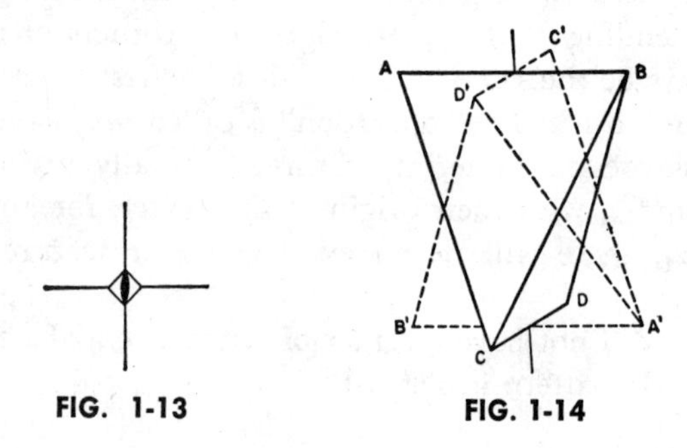

FIG. 1-13 FIG. 1-14

indicating the 4-fold rotatory-inversion axis, the solid almond indicating the 2-fold rotation axis coincident with it, and the straight lines indicating the two mirror planes.

The mirror plane describes symmetry with respect to a plane, the rotation axis describes symmetry with respect to a line, and the center of symmetry describes symmetry with respect to a point. The inversion axis describes symmetry with respect to an axis plus a point. The operation of rotatory inversion is a combined operation, and if you try to separate its parts, you destroy its character completely. Note how it operates on the disphenoid (Fig. 1-14). After a rotation of 90° around the vertical axis, *A′* of

* Note that the symmetry *operation* is the *act* that brings an object into a position indistinguishable from its former position. The symmetry *element* is the *representation* of these acts.

the rotated object (formerly point D) is now related to A of the original object by inversion through the center, and similarly for B' and B, C' and C, D' and D, as well as for the edges and faces. Thus the *combined rotation and inversion* brings D into A via A'. Neither one acting alone can bring D into coincidence with A. Therefore the disphenoid does not have a 4-fold axis and it does not have a center of symmetry, but it does have a 4-fold axis of rotatory inversion. It has other axes too. (See p. 67.)

In the case of the three-circle pattern, the table, the brick, and the sphenoid, the symmetry operations we performed always left one point unmoved: the center of the object. The group of operations on an object that leave one point in the object unmoved during the operations as well as keeping distances between all points unchanged, as in a rigid body, is called the *point group*. Except in Appendix II, we will not need to consider any other kinds of symmetry operations.

Every symmetry operation must, of course, be consistent with every other symmetry operation in the point group. This means that while certain assemblages of symmetry operations can make up a point group, others cannot.

For example, a 3-fold axis cannot have a single mirror plane parallel to it because the axis requires that every feature of the object, including its mirror plane, be repeated three times in a complete revolution around the axis. Another way of expressing the same thing is to say that, if you design an object with a 3-fold axis of symmetry and a mirror plane parallel to this 3-fold axis, you will find that the other two symmetry planes exist at $\pm 120°$ to the first.

This interdependence of the symmetry operations in a point group can be explored more easily when we have a diagrammatic way of looking at three-dimensional arrangements of the elements that represent them. This diagrammatic method will be developed in Chapter 4.

In Chapter 2 we will find that not all possible point groups can be used to describe the symmetry of arrangement of atomic planes in crystals. The homogeneous periodic array of the crystal structure can only have certain kinds of symmetry, and this limits the number of possible *crystallographic point groups* to 32.

* * *

If you want to practice determining the point group of various objects, here are a few suggestions. (Answers are given at the end of Chapter 2.)

What are the symmetry elements of:

1. The human body, disregarding the insides?
2. A shoe?
3. A cube?
4. A box kite?
5. A pinwheel constructed by cutting the diagonals of a square nearly to the center and bringing one corner of each triangular wing to the center point to be impaled on a pin? (See Fig. 1-15.)

FIG. 1-15

2 *Symmetry in Crystals*

A crystal is a solid composed of atoms arranged in an orderly repetitive array. That was the way Chapter 1 began, and in that chapter we explored symmetry, beginning with the symmetry of the orderly arrangement of some of the atoms in the crystal calcite and progressing to the symmetry of a number of familiar objects.

The fact that in a crystal the pattern of arrangement of the atoms is repeated in all directions puts certain restrictions on the kinds of symmetry crystals can have. In this chapter we will see what these restrictions are.

Nearly all crystals actually have various kinds of imperfections scattered here and there in their structure which spoil their unending perfection, and in all crystals the atoms are vibrating with thermal motion, some more, some less, according to the temperature of the crystal and the strength of attraction of the atoms for each other. However, these are minor modifications of the ideal structures of our models. In the models the atoms are shown as balls, although we know that in some cases an atom will share some of its electrons with certain adjoining atoms so that it should more properly be shown as having protuberances toward some of its neighbors. But if we regard each ball as indicating the average position of the center of each atom in the perfect parts of the crystal, the models will serve very well.

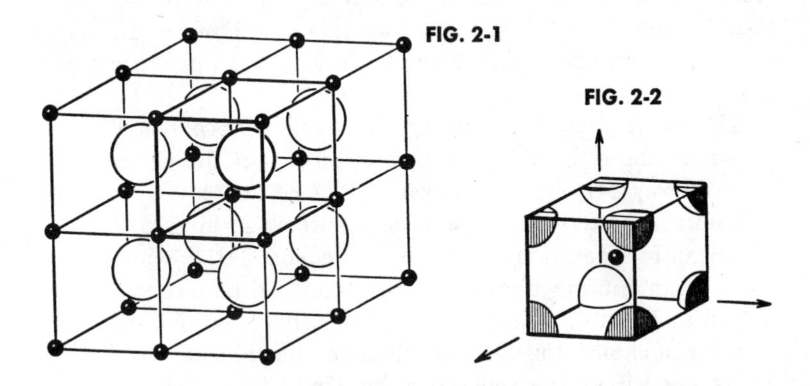
FIG. 2-1

FIG. 2-2

Fig. 2-1 shows the pattern of the arrangement of the atoms in cesium chloride. Lines connecting the positions occupied by the cesium atoms help to guide the eye. They divide the crystal into cube-shaped cells, each of which has a chlorine atom in the center. If these lines had been drawn connecting the chlorine atoms to each other, each cell would have had a cesium atom in the center, as shown in Fig. 2-2. In each corner is one-eighth of a chlorine atom, since eight cubic cells share this atom where they meet at the corner. The cell thus contains one cesium atom and one chlorine atom (eight eighths), in accordance with the formula of the substance, CsCl. The repetition of this unit of pattern results in the orderly structure of the cesium chloride crystal. If we translate (not rotate) the unit cell shown in Fig. 2-2 in the direction of any one of the arrows (the directions of its edges) by a distance equal to the length of its edge, it will occupy exactly the position of the neighboring cell that is just like it. This repetition, which may be regarded as resulting from the operation of translation alone, is sometimes called translational symmetry.

The smallest pattern unit of the arrangement of atoms in a crystal from which the whole crystal can be built up by (mentally) translating the unit parallel to itself in the direction of its edges by distances equal to the lengths of its edges is called the *primitive unit cell* of the crystal. Although an infinite variety of such cells can be chosen, each with the same volume as the others, it is commonly easy to choose one that is preferable to the others, usually one that exhibits the maximum symmetry. In some cases we have reasons for choosing a unit cell larger than the primitive cell; these cases will be explored later.

The guide lines outlining the unit cell need not have been drawn to the centers of atoms. The corners of the cubes could lie half way between neighboring chlorine atoms, and contain one unit of pattern just as well, or anywhere so long as the size, shape and orientation of the units remained the same. Since there is an infinite number of points in the pattern-unit cube shown in Fig. 2-2, we can truthfully say that there is an infinite number of choices for the unit cell of cesium chloride. Ordinarily we choose either the cell shown in Fig. 2-1 or that shown in Fig. 2-2.

These unit cells may be regarded as the "building blocks" that make up the crystal, each one indistinguishable from the next. They must have a shape that will enable them to fit together to fill space solidly. A shape such as the disphenoid, for example, would not fulfill this condition. What shapes would?

Clearly, space can be filled solidly with cubic cells that are exactly alike. What if each cell or pattern-unit is longer in one direction than it is in the other two? These cells still fill space solidly, as shown in Fig. 2-3, and so will the cells that have different dimensions along the *a*, *b*, and *c* directions of Fig. 2-4, like bricks, so long as each is exactly like all the others. In Figs. 2-1, 2-3, and 2-4, the angles of the cells are all 90°. But the space-filling property of the cells is not spoiled if the whole structure is skewed, as in Fig. 2-5.

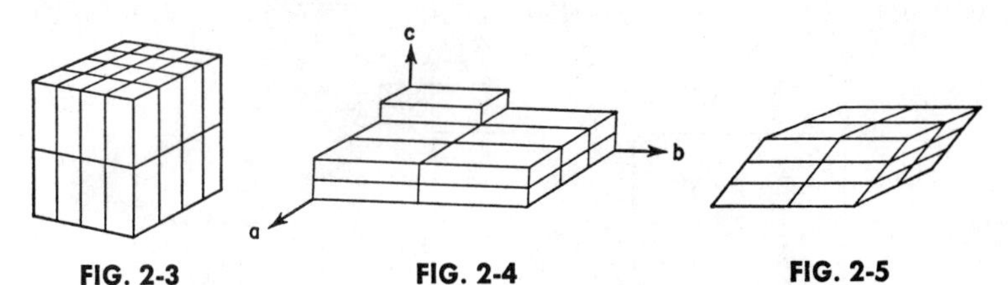

FIG. 2-3 FIG. 2-4 FIG. 2-5

So far, all of these cells have had their opposite faces parallel. The general term for a body with parallel surfaces is a parallelepiped. Can we fill space solidly with any other kind of cell?

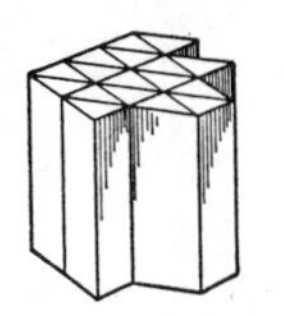

FIG. 2-6

FIG. 2-7

In Fig. 2-6 triangular prisms are packed together to fill space, but we find that each one is not derived from its neighbor by straight translation in a direction parallel to one of its edges, so they will not serve as unit cells. However, a pair of such triangular prisms makes a proper unit cell, as we can see by looking down on the top of the prisms (Fig. 2-7).

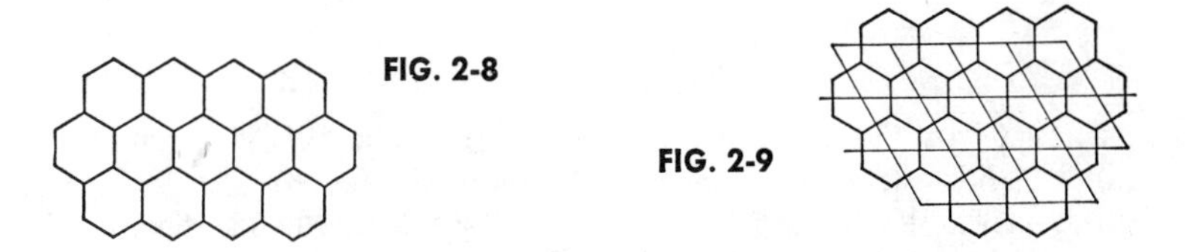

FIG. 2-8 FIG. 2-9

In Fig. 2-8 we are looking down on the tops of a number of hexagonal prisms packed together to fill space solidly. Tiled floors are often made this way. But here again we find that each hexagonal prism is not derived from its neighbor by straight translation parallel to one of its edges.

However, we can choose a unit of pattern that is so derived, and this is shown in Fig. 2-9. Since five-sided prisms and more-than-six-sided prisms cannot be packed together to fill space, we will not be concerned with them at all. We are, in fact, left with

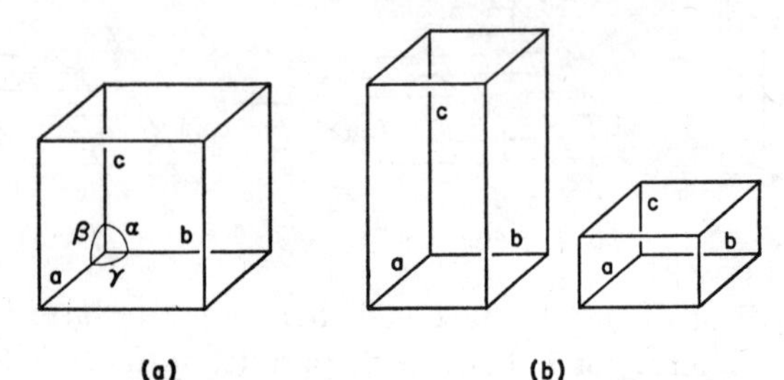

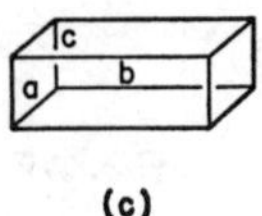

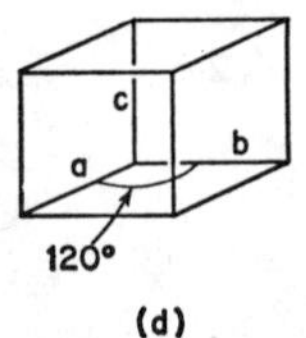

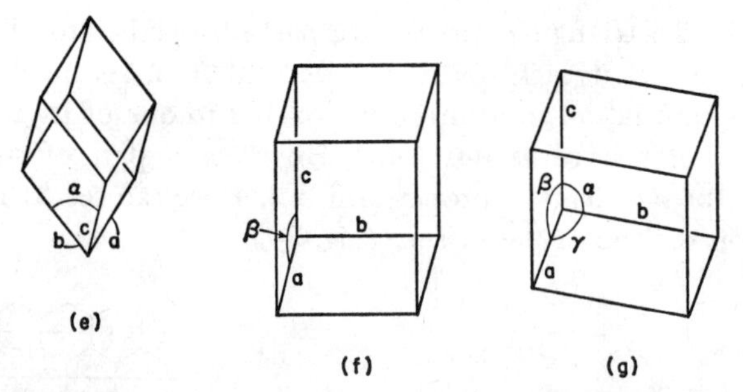

FIG. 2-10 (a) Cubic cell: $a = b = c$, $\alpha = \beta = \gamma = 90°$. (b) Tetragonal cell: $a = b \neq c$ (c is either longer or shorter than a and b), $\alpha = \beta = \gamma = 90°$. (c) Orthorhombic cell: $a \neq b \neq c$, $\alpha = \beta = \gamma = 90°$. (d) Hexagonal cell: $a = b \neq c$, $\alpha = \beta = 90°$, $\gamma = 120°$ (as in Fig. 2-9). (e) Rhombohedral cell: $a = b = c$, $\alpha = \beta = \gamma \neq 90°$. (f) Monoclinic cell: $a \neq b \neq c$, $\alpha = \gamma = 90°$, $\beta > 90°$. (g) Triclinic cell: $a \neq b \neq c$, $\alpha \neq \beta \neq \gamma$.

just the parallelepipeds like those in Figs. 2-1, 2-3, 2-4, 2-5, 2-7, and 2-9. When we classify their possible shapes systematically, we find seven types that form the crystal building blocks of the seven different "systems" of crystals. These are shown in Fig. 2-10. We will call the lengths of the three edges a, b, and c and the angles between each pair of edges α, β, and γ (alpha, beta, and gamma), in each case using the Greek letter whose Roman equivalent is missing from the pair of edges (Fig. 2-10a).

Some possible parallelepipeds have been omitted because they were equivalent to others that were included. For example, in Fig. 2-11 we see $a = b \neq c$, $\alpha = \beta = 90°$, $\gamma \neq 90°$, but looking

FIG. 2-11 FIG. 2-12

down on these prisms from the top, as in Fig. 2-12, we see that the same pattern can be considered as made up of orthorhombic cells that have the advantage of being orthogonal (i.e., $\alpha = \beta = \gamma = 90°$). The new a and b are not shown in Fig. 2-12.

The fact that crystals are made up of such "building blocks" or repeat units of some three-dimensional pattern has been known a very long time. Crystals like those shown in Plate II and in Plate III (1) occur in rocks and can be pried out—without damage, if the job is skillfully done—or the weathering of the rock loosens them and they lie in the soil, where their smooth plane faces reflect the light and attract the attention of children and naturalists who explore unlikely places.

It is in the nature of naturalists to describe in detail the odd things that they find, and the early naturalists wrote full descriptions of the crystals they found, recording carefully the angles between their faces. One of them (p. 70) noticed the special angular relationships between the faces of a crystal that led him to the conclusion that crystals were made up of "building blocks." We can repeat his discovery by looking at the topaz crystal in Plate III (1) and measuring the angles that its small side-faces make with the long side-face of the crystal. In Fig.

2-13, where the outline of the crystal is traced from the photograph, we find that the angles indicated are 24.5° and 42.3°. The tangents of these angles are 0.456 and 0.910, or 1×0.456 and 2×0.456.

How can we explain the fact that these tangents are related by small whole numbers? Let us suppose that we have a stack of bricks, as shown in Fig. 2-14, with a board placed against the

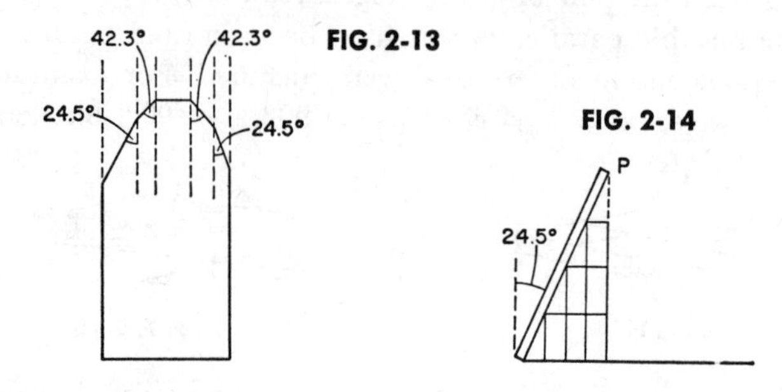

FIG. 2-13

FIG. 2-14

edges of the bricks at an angle of 24.5° to the vertical. The tangent of this angle is 0.456.

To the right of point *P* we wish to stack the bricks so that a board laid against them will be at an angle to the vertical of 42.3°, whose tangent is 2×0.456. How do we stack the bricks to accomplish this? Get your bricks stacked before you read further.

The solution is shown in Fig. 2-15. By setting the bricks back by two brick-widths in each step we have created a slope making an angle with the vertical of 42.3° whose tangent is just twice that of the angle 24.5°. Of course, if the bricks were set back by three brick-widths in each step, the tangent would be three times the tangent of 24.5° and would correspond to an angle of 53.8°. It was this small-whole-number relationship between the tangents of the interfacial angles of crystals that convinced the early naturalists that crystals were made up of building blocks.

One of the early crystallographers was J. B. L. de Romé de l'Isle, who published an *Essai de cristallographie* in Paris in 1772. He showed that the various faces of a crystal can be derived from what he called a "primitive form" by replacing its edges and solid angles by planes. The Abbé René Just Haüy* (pronounced "Howie") carried this idea further in his *Essai d'une théorie sur la structure des crystaux,* published in Paris in 1784, enunciating clearly the whole-number relationship which we have just rediscovered.

Of course, the Abbé Haüy was bothered by the fact that the edges of the building blocks could not be detected at the faces of the crystal, even with the strongest microscope. Nonetheless, the small-whole-number relationship provided such convincing evidence of the existence of the repetitive structural units that he concluded that those building blocks just must be extremely small, so small that they did not even make the crystal faces look frosty the way the surface of ground glass looks because it is slightly roughened.

How big would the irregularities have to be in order to scatter the light the way ground glass does? They would have to be somewhere near the dimensions of the wavelength of light, about 5×10^{-5} cm, although this figure was not known until the first decade of the nineteenth century. So the naturalists following Haüy knew that the pattern units of structure in crystals were smaller than 10^{-5} cm, and they knew the units' shapes quite accurately from their measurements of the interfacial angles on crystals they had found. But they had no way of knowing what the pattern within the unit cell consisted of, nor could they tell the cell size.

We can now determine with great accuracy the dimensions of the unit cells of crystals, and the arrangement of atoms in the cell is known for a very large number of substances. These results are obtained by the diffraction of x-rays from the orderly three-

* René Just Haüy was born at St. Just, just north of Paris, in 1743. His parents could not afford to send him to college, but friends who recognized the young man's ability made it possible for him to receive advanced education. He had become interested in botany when an accident made him become a crystallographer. The account of this accident must be postponed until Chapter 7. Imprisoned during the French Revolution, he later became a professor of mineralogy under Napoleon's reign. However, with a change of government in 1814 he was discharged and was again a very poor man when he died in 1822.

dimensional array of atoms in the crystal. When a beam of x-rays enters a crystal, each atom in the crystal scatters x-rays in all directions. X-rays, like light, can be represented as waves (Fig.

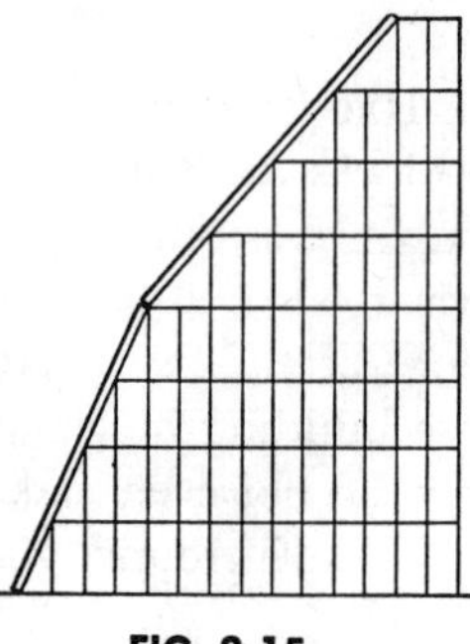

FIG. 2-15

2-16), and as in the case of visible light, the distance from crest to crest (or any point to the next similar point) of the wave is called the wavelength, usually indicated by lambda, λ. In visible

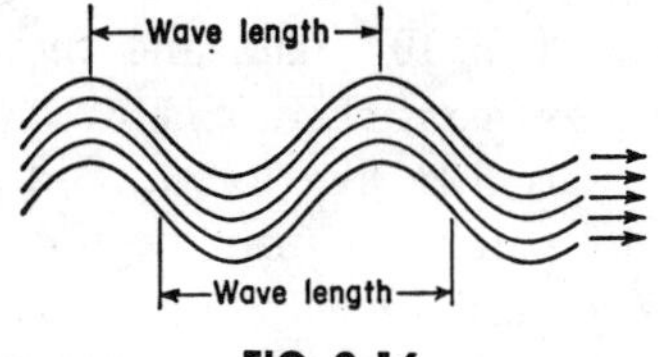

FIG. 2-16

light the length of these waves ranges from about 4×10^{-5} cm to about 8×10^{-5} cm, but in x-rays the wavelengths range from about 0.1×10^{-8} cm to about 10×10^{-8} cm, or about 1/1000 of the length of those of visible light.

Fig. 2-17 shows what happens when a beam of x-rays meets a crystal. Here the horizontal lines represent two layers of atoms in the crystal (like two of the sheets of cesium atoms in Fig. 2-1), seen edge on. To simplify the picture, only one atom in each layer is shown. The distance between the two layers is labeled *d*. A beam of x-rays represented by rays *A* and *B* is incident upon the surface of the crystal. Its wavelength, λ, is indicated by a marker at the position of each wave crest. The two rays, *A* and *B*, of this beam meet the atoms in the two layers and bounce off in all directions, several of which are shown, the

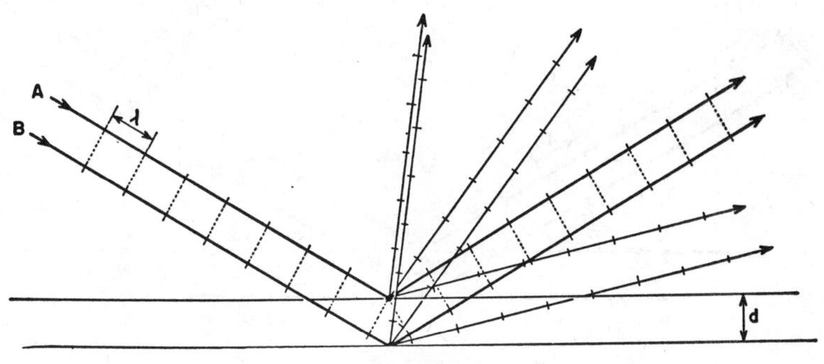

FIG. 2-17

wavelength markers being properly measured off on every ray. The two rays start off with their crests and troughs in step, so to speak, or "in phase." However, the fact that ray *B*, which penetrates more deeply into the crystal, travels a longer path means that it gets behind. Only if it gets behind by exactly one wavelength (or exactly two, three, four, or any whole number of wavelengths) will it be in phase with the other ray again, and this, as you can see from Fig. 2-17, will only happen in certain directions of the scattered rays. In these directions the beam leaving the crystal toward the right will be strong, whereas in directions where the waves are not in phase it will be weak.

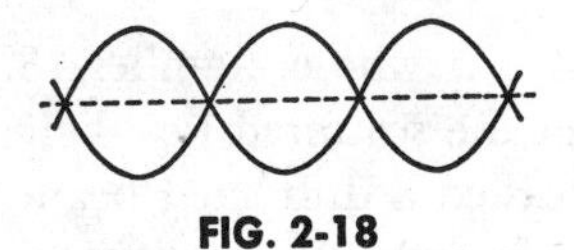

FIG. 2-18

Fig. 2-18 is a diagram of two waves that are exactly out of phase: one is retarded with respect to the other by half a wavelength, so that the crest of one is in line with the trough of the other. You can see that the sum of their displacements from the middle line is zero: such waves would "cancel" each other and the resulting intensity would be zero.

The diagram in Fig. 2-19 is a geometrical analysis of the conditions necessary for a strong beam to occur. In this diagram, the darkened section of the path of ray B is the extra distance it

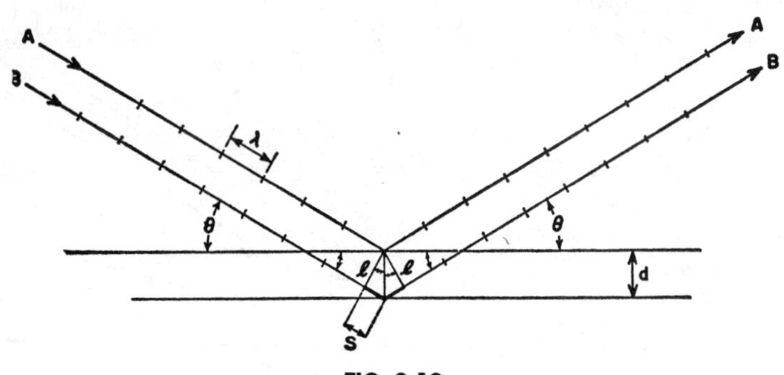

FIG. 2-19

travels, over and above that traveled by ray A, which is compared by drawing lines, l, perpendicular to both rays. As we have just seen, this extra path has to be a whole number of wavelengths long in order for the ray B to be in phase with ray A on leaving the crystal. All the arc-marked angles in Fig. 2-19 are θ, and in each of the little right triangles, $\sin\theta = \frac{s}{d}$. The extra path length is $2s$ long, or $2d \sin\theta$ long, and since this value must equal some whole number of wavelengths for a strong beam, the condition for the strong beam is that

$$n\lambda = 2d \sin\theta \text{ (the Bragg equation)}$$

where n is any positive whole number. Such a beam, whose intensity results from the scattered rays being in phase in a particular direction, is called a diffracted beam, and the process we have been describing is called x-ray diffraction.*

When the Bragg equation is satisfied, the angle that the diffracted beam makes with the diffracting planes of atoms is the same as that which the incident beam makes with the same

* The equation $n\lambda = 2d\sin\theta$ is known as *the Bragg equation,* and θ is known as *the Bragg angle.* Sir Lawrence Bragg, working in Cambridge, England, first derived it and used it in the determination of the structures of crystals in 1912.

plane. In this way, but only in this way, diffraction is like reflection of light from a mirror, in which "the angle of incidence is equal to the angle of reflection." * Because of this, one speaks loosely of x-ray "reflection" from a crystal, the "reflected beam," and so forth. This terminology is so common that it will be used in this book, but we must remember as we use it that we really mean diffraction.

Unlike a two-dimensional diffraction grating, the three-dimensional diffraction grating will not diffract monochromatic light (p. 79) falling on it from any angle. The great power of x-ray diffraction as a tool for studying crystals lies in the fact that the diffracted beam *can only occur* when a beam of a particular wavelength, λ, meets like planes of atoms in the crystal that are a distance d apart and meets these planes at precisely that angle, θ, whose value will satisfy the Bragg equation for the particular wavelength of x-rays used and the particular interplanar spacing of the atoms.

Clearly, if we know the wavelength of the x-rays we are using and the angle at which a strong beam is diffracted from the crystal, we can easily calculate the spacing between like planes of atoms in the crystal. This is how we now know the dimensions of the unit cells of crystals. The edge of the cubic cell of cesium chloride is 4.12×10^{-8} cm long. How many atoms are there per cubic centimeter in a crystal of cesium chloride? (The answer appears at the end of this chapter.)

Since the unit cells of most substances have dimensions that range from 2.5×10^{-8} to 10×10^{-8} cm, a special unit is used which is equal to 10^{-8} cm. It is called the Ångström Unit.† (The letter Å follows Z in the Swedish alphabet, which also

* By convention, the glancing angle of the x-rays is measured between the incident beam and the "reflecting" planes of atoms. In optical reflection the convention is to measure the angle of incidence between the incident beam and the normal to the reflecting planes, as in Chapter 12.

† Anders Jonas Ångström, born in 1814, was the Swedish physicist who first deduced on theoretical grounds that an incandescent gas emits light of the same wavelength as that of the light that it selectively absorbs. This is a very important fact in spectroscopy, the study of the absorption and emission or radiation by matter. In 1862 Ångström announced his discovery of the presence of hydrogen in the sun's atmosphere, based on spectroscopic evidence.

contains the letter A as the first letter. The letter Å is pronounced "oh.")

It was a young instructor in the university at Munich in 1912 who first thought of using x-rays in this way to measure the distance between layers of atoms in crystals. His name was Max von Laue. At that time, physicists were not yet sure what x-rays were, although they knew you could cast shadow-pictures of the bones of the hand with them. Wilhelm Röntgen, who had discovered them in 1895, had demonstrated this. Sir J. J. Thomson and others thought that x-rays were like visible light, but of much shorter wavelength.

It was generally known at this time that if very fine parallel lines are scratched on a reflecting surface with a regular spacing between them and a beam of visible light of a given color (i.e., wavelength) is directed onto the surface, diffracted beams will come off at special angles where the light waves are in phase. Such a ruled surface is called a diffraction grating. When Max von Laue learned from his friend Paul Ewald that crystals are made up of very small "building blocks" arranged in layers, it occurred to him that this small-scale three-dimensional grating might be a good diffraction grating for x-rays if they were in fact the "small-scale light rays" that some thought they might be. He therefore suggested to his two young assistants, Friedrich and Knipping, that they allow a narrow beam of x-rays to fall on a crystal and place a film (protected from visible light by black paper) nearby. When the film was developed, it showed distinct spots, proving that the x-rays scattered from the crystal did indeed come off in distinct beams at certain angles, that is, were diffracted by the "three-dimensional grating" of the orderly structure of the crystal. A photograph taken by letting a narrow beam of x-rays fall on a tungsten crystal and reflect back onto the film is shown in Plate III (3). It is called a Laue photograph. Ever since this historic experiment, performed in 1912, people have been learning about the structure of crystals by examining the spots made on films by x-rays diffracted from the crystal. Knowledge of the wavelength, λ, responsible for a particular spot and the angle, θ, of reflection makes possible the calculation of the distance, d, between atomic planes. Careful measurement of the intensities of the spots enables us to check whether the atoms are arranged in one way or another.

To show how the intensity of the diffracted beam is affected by the arrangement of the atoms, we can take the cesium chloride structure (Fig. 2-1) as an example. If a beam of x-rays scattered in a given direction from the top layer of cesium atoms is just $n\lambda$ ahead of the beam from the next layer, the two beams will be in phase and, combined with the similarly in-phase beams from lower layers, will result in a strong diffracted beam. But the same incoming beam of x-rays will also be scattered by the layers of chlorine atoms which are just half way between the layers of cesium atoms. The extra path length will be just half that for the cesium atoms (Fig. 2-20), and the resulting diffracted

FIG. 2-20

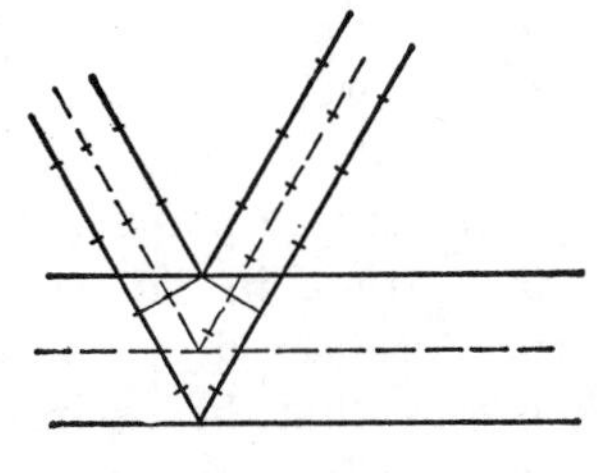

FIG. 2-21

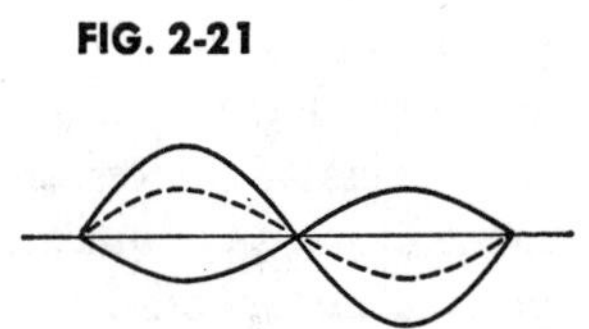

rays from the chlorine atoms will be just exactly out of phase with those from the cesium atoms. If the two kinds of atoms had the same scattering power for x-rays, the amplitude (half the distance from the crest to the trough bottom) of their scattered waves would be the same and they would cancel each other, as in Fig. 2-18. However, since the cesium and chlorine atoms have different x-ray scattering powers which depend on their total numbers of electrons (55 in cesium and 17 in chlorine), the diffracted wave is only *weakened* by the out-of-phase contribution from the chlorine atoms, as in Fig. 2-21, where the broken line is the sum of the two curves. Thus the fact that this reflection has an intensity which can be accounted for by the difference in scattering power of the two kinds of atoms tells us where the atoms are. Usually, the intensity of many reflections from many differently oriented layers of atoms in a crystal are needed for the determination of its structure. In the case of complicated structures, many hundreds of such measurements are made and the analysis of the structure may take years of work, even with present-day computing machines.

A cubic crystal in which the atom or group of atoms at the center of the cubic cell is indistinguishable from that at the corners has a "body-centered cubic cell." In order to discuss this crystal, it will be useful to have a name for a repetitive array of unit cells like those in Figs. 2-3, 2-4, and 2-5. Since these frameworks are lattices in three dimensions, they are called *space lattices.* Their cross points are called *lattice points.* If we remember that such a lattice goes on for so many millions of cells in a crystal that it can be considered infinite in extent when we are thinking of parts of it, we see that every lattice point is indistinguishable from every other lattice point. If you could sit, with no dimensions, at any lattice point, your surroundings would look the same as they would if you sat at any other lattice point. In Fig. 2-1 a lattice has been drawn with its corners at the cesium atoms, and it is clear that the environment of each cesium atom is indistinguishable from that of the other cesium atoms. In Fig. 2-22 the same lattice has been trans-

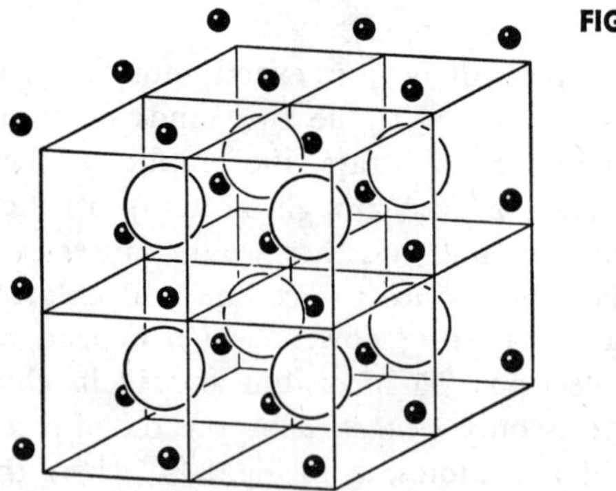

FIG. 2-22

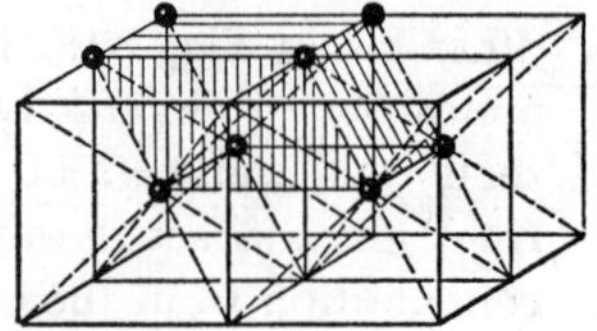

FIG. 2-23

lated, relative to the cesium chloride structure, but each lattice point is still indistinguishable from every other lattice point: each has a cesium atom a short distance up to the left of it, for example.

To return, now, to the body-centered cubic cell, it was defined as having an atom or group of atoms at the center of the cubic unit cell, indistinguishable from that at the corners. In other words, it has a lattice point at its center as well as at its corners. If this is so, it must be that a smaller unit cell could be chosen. Such a cell is shown, shaded, in Fig. 2-23. Comparison of its edge lengths and angles with those in Fig. 2-10 will show that it is triclinic. Without making any measurements, we can see that its volume must be just half that of the cube-shaped cell, since it contains only one lattice point (remembering how the shared points at the corners are counted), whereas the cubic cell contains two. A cell that contains only one lattice point is a primitive cell,* and a lattice made up of such cells is called a primitive lattice. A primitive unit cell can always be chosen for any crystal structure, but sometimes, as in the case of the body-centered cubic cell, we prefer to choose a cell that is double the size of the primitive cell (or more than double) because its symmetry is that of the lattice.

For the same reason, both body-centered and face-centered unit cells are used in other systems besides the cubic system. When all these systems are counted, we find that there are 14 space lattices. Cells from each of these are shown in Fig. 2-24, where the lattice points are shown as black dots. Seven of these are the primitive cells we started with.

The proof that the number of possible parallelepipedal networks of points is 14 was given by Auguste Bravais† in 1848, and the 14 space lattices are therefore sometimes called the Bravais lattices. They had been derived six years earlier by Frankenheim‡, but Frankenheim did not rigorously prove his

* Another primitive cell has rhombohedral symmetry.

† Auguste Bravais, born in southern France in 1811, was a naval officer as a young man. Later, through his interest in various aspects of mathematical physics, including astronomy, he held the Chair of Physics at the Ecole Polytechnique in Paris.

‡ Moritz Ludwig Frankenheim, Professor of Natural Philosophy at Breslau.

derivation, and in fact made the mistake of including a fifteenth lattice which, Bravais showed, was the same as one of the others.

In 1962 a Commemoration Meeting was held in Munich on the fiftieth anniversary of the Laue experiment, which had been performed in Munich. When plans for the meeting were first made, von Laue was alive and active, but he was killed in 1960

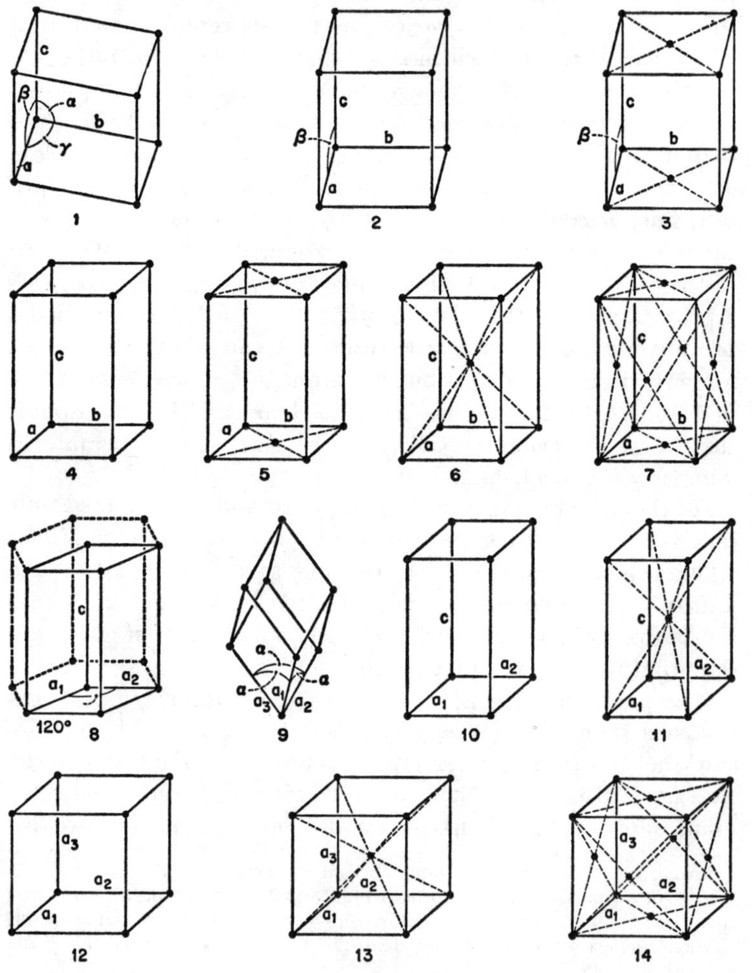

FIG. 2-24

when the car he was driving collided with a motorcycle. Paul Ewald had now become Professor Emeritus Ewald of the Polytechnic Institute of Brooklyn, President of the International Union of Crystallography (a scientific union, not a labor union) and Editor-in-Chief of Acta Crystallographica, the international crystallographic journal. For the Munich celebration he undertook the enormous task of preparing a volume entitled "Fifty Years of X-ray Diffraction." Much of it is written by Professor Ewald himself, but some 35 or more other crystallographers have contributed their personal reminiscences and accounts of the development of the science.

This book is richly rewarding reading. In his autobiography, von Laue gives such personal items as "As sources for the primary current we used chromic acid and Bunsen elements, as the apartments of our parents were not connected to the city's electric power. Many a hole did the required chemicals burn in our clothes." "I was plunged into deep thought as I walked home along Leopoldstrasse just after Friedrich showed me this picture. Not far from my own apartment at Bismarckstrasse 22, just in front of the house at Siegfriedstrasse 10, the idea for a mathematical explanation of the phenomenon came to me."

Professor Ewald writes of the days when he was a graduate student at Sommerfeld's Institute in Munich, "Even more efficiently and informally than at the Institute an exchange of views and seminar-like consultation on any subject connected with physics took place in the Café Lutz in the Hofgarten, when the weather permitted under the shade of the chestnut trees, and otherwise indoors. This was the general rallying point of physicists after lunch for a cup of coffee and the tempting cakes. Once these were consumed, the conversation which might until then have dealt with some problem in general terms, could at once be followed up with diagrams and calculations performed with pencil on the white smooth marble tops of the Café tables—much to the dislike of the waitresses who had to scrub the tables clean afterwards." Concerning the man who got the Nobel Prize for the discovery of x-rays, he writes, "No need to say that Röntgen never came to this informal meeting—nor even to the regularly scheduled Physics Colloquium; he was dominated by a shyness that made him evade personal contact wherever he could."

As indicated in the footnote on page 20, it was W. H. and W. L. Bragg, father and son, who originated the science of crystal structure analysis by X-ray diffraction. By 1914, only two years after the Laue experiment, they had succeeded in determining the arrangements of atoms in sodium chloride, potassium chloride, potassium bromide, potassium iodide, zincblende, fluorite, calcite, pyrite, and diamond.

* * *

Problem: In a face-centered cubic lattice, each cubic unit cell has a lattice point in the center of each face in addition to those at the corners. (1) Draw a face-centered cubic cell and outline a primitive cell, as we did in Fig. 2-23 for the body-centered cubic case. (2) What is the space lattice of the primitive cell? (3) How much larger is the f.c.c. cell than the primitive cell? (Answers are given at the end of Chapter 3.)

* * *

Answer to question in Chapter 2: In a cubic unit cell of cesium chloride, 4.12×10^{-8} cm on an edge, there are two atoms (Fig. 2-2). How many of these unit cells are there in a cube 1 cm on an edge? Along one edge there are $1/4.12 \times 10^{-8} = 0.243 \times 10^{8} = 2.43 \times 10^{7}$. Along one face there are $(2.43 \times 10^{7})^{2}$, and in the entire cube there are $(2.43 \times 10^{7})^{3}$, or 14.35×10^{21} cells with two atoms in each cell. Therefore there are 2.9×10^{22} atoms per cc in cesium chloride.

* * *

Answers to questions at the end of Chapter 1

1. One vertical mirror plane halfway between the ears. Actually even the exterior of most people does not obey in detail such a symmetry element. An amusing photographic experiment is to cut a photograph of a face in two and replace the left half with the mirror image of the right half (printed by reversing the negative). Even when skillfully produced, the finished face does not fully resemble the person, and most people have difficulty determining what change has been made. Some people are very unsymmetrical, and their portraits would be almost unrecognizable after such treatment.
2. None.
3. Nine planes, three 4-fold axes, four 3-fold axes, six 2-fold axes, and a center. (See Fig. 2.25.)
4. One 4-fold axis, four 2-fold axes, five planes, and a center. (See Fig. 2-26.)
5. One 4-fold axis normal to the paper.

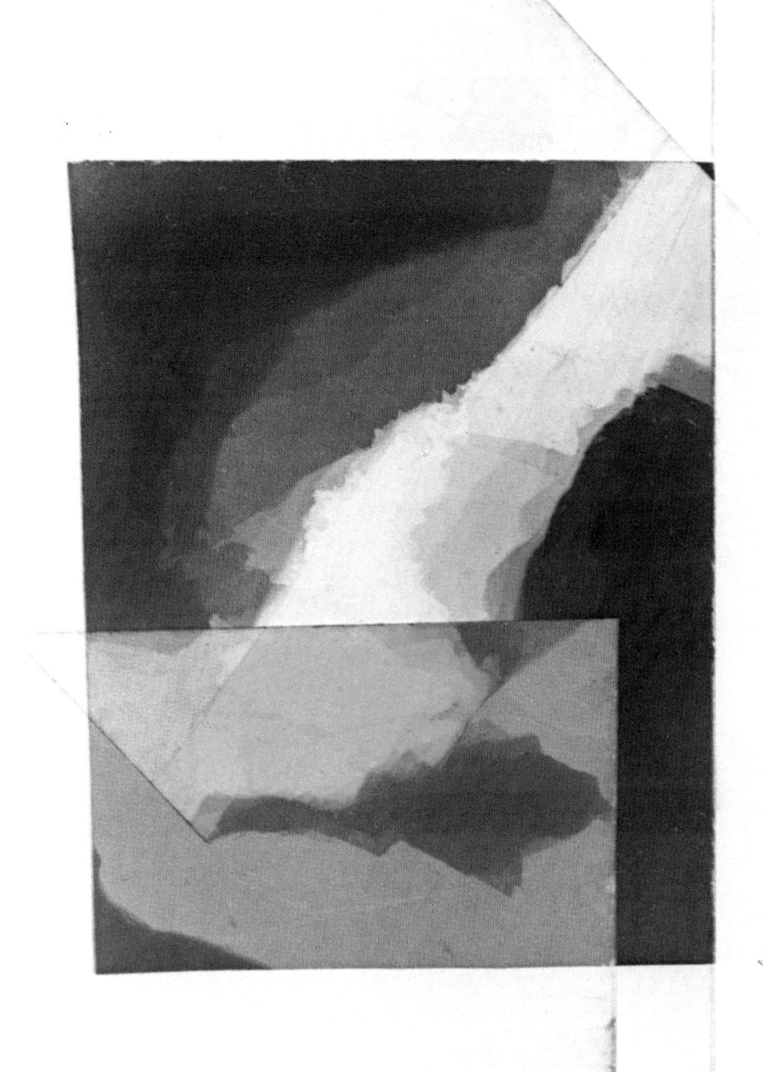

PLATE I Mica on mica between crossed polarizers.

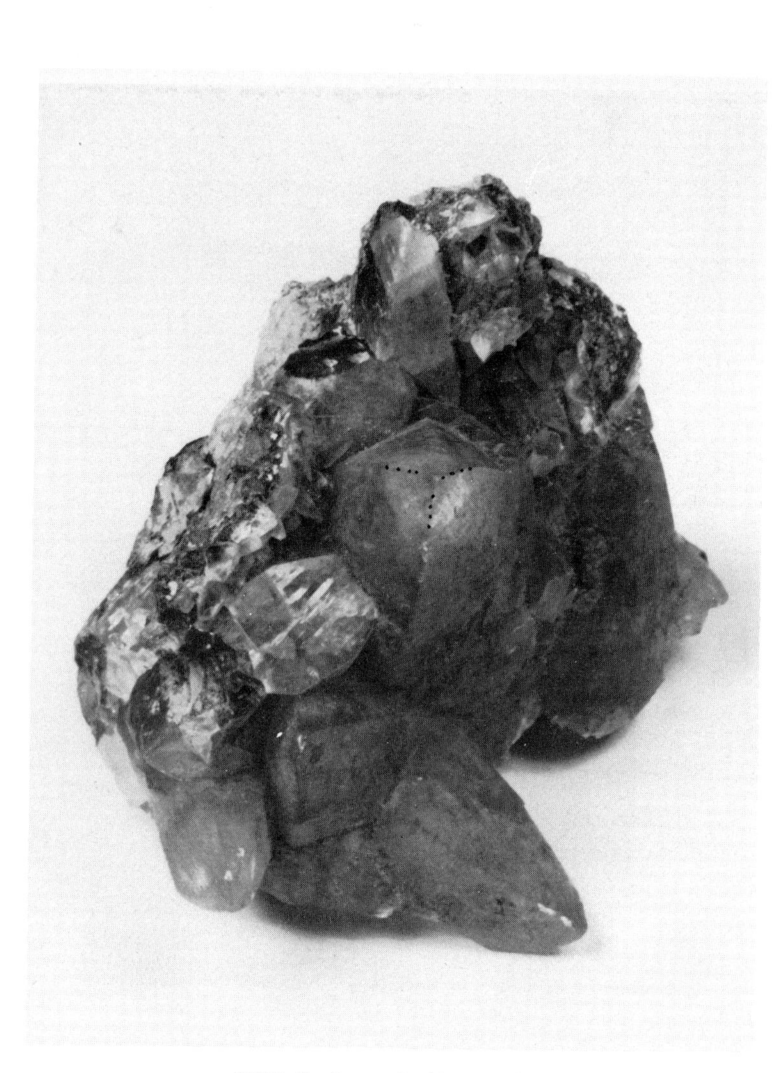

PLATE II Group of calcite crystals.

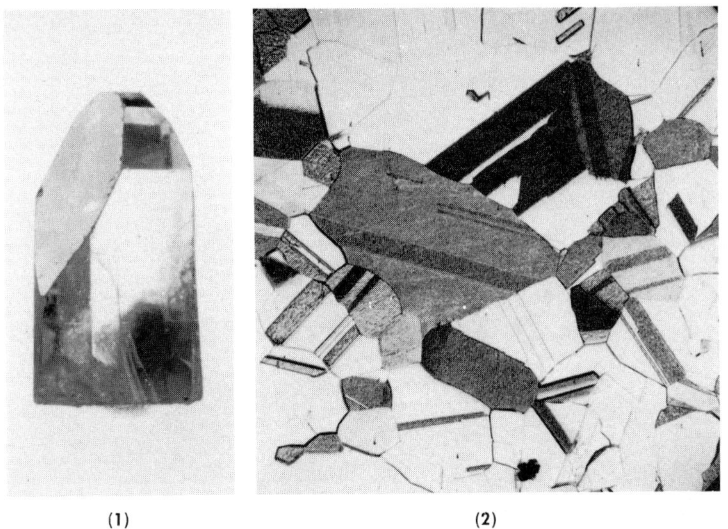

(1) (2)

(3)

PLATE III (1) Topaz crystal. (2) Polycrystalline brass showing twinning on (111) polished and etched. (Photomicrograph by F. G. Foster, from *Crystal Orientation Manual*, by Elizabeth A. Wood, by permission of Columbia University Press). (3) Back reflection Laue photograph.

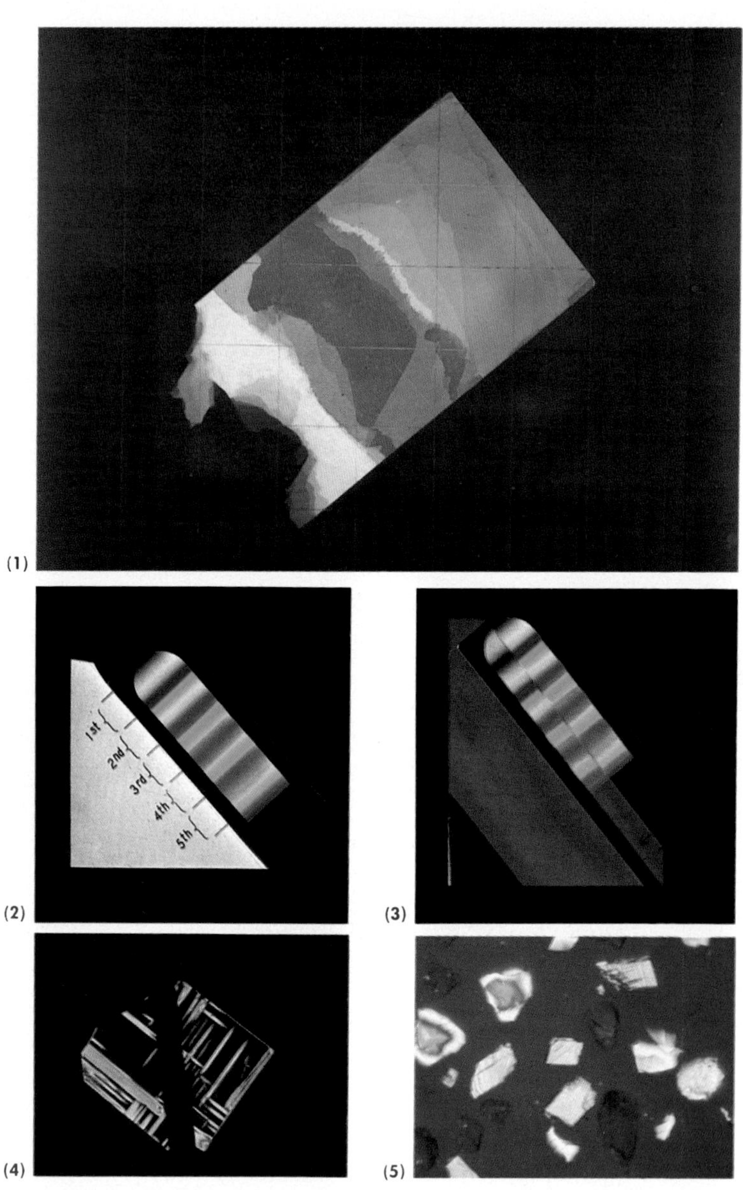

PLATE IV (1) Mica between crossed polarizers (lines show polarization directions of Polarizer and Analyzer). (2) Quartz wedge between crossed polarizers: successive orders of interference colors indicated. (3) Quartz wedge and Scotch tape between crossed polarizers, showing that the length direction of the tape is the vibration direction of the slow ray. (4) Barium titanate between crossed polarizers. (5) Calcite and quartz fragments between crossed polarizers (Photomicrograph courtesy of ERNST LEITZ *GMBH*, WETZLAR).

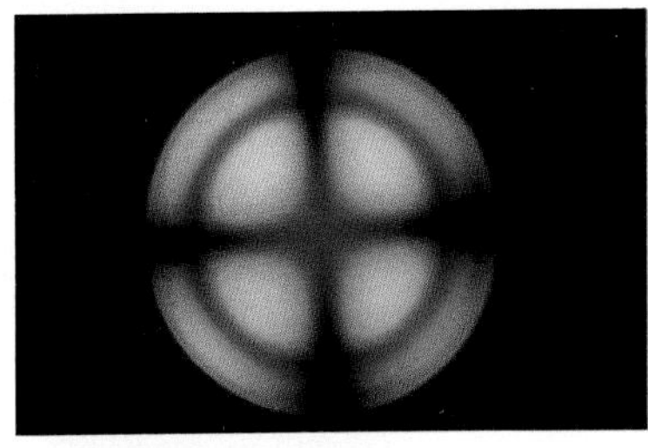

(1a)

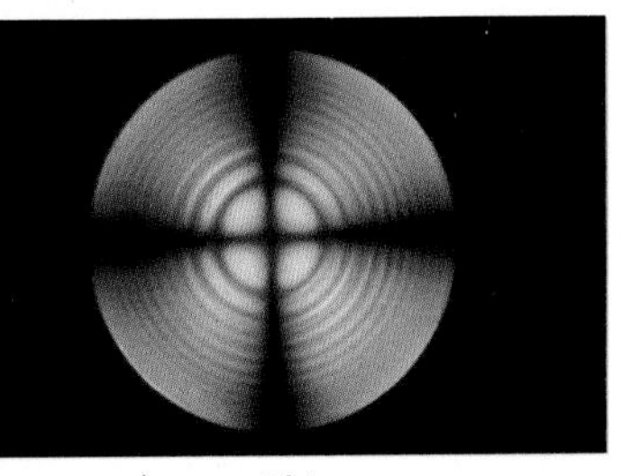

(1b)

(2)

(1) Uniaxial interference figures with the polarizing microscope. (a) Quartz. (b) Calcite. (2) Uniaxial interference figure with gypsum plate (quartz). (Photomicrographs courtesy of ERNST LEITZ *GMBH*, WETZLAR).

(3) Uniaxial interference figure without the microscope (guanidinium aluminum sulfate hexahydrate). The specimen is shown actual size. Note shift of interference bands where thickness changes at surface step.

PLATE V

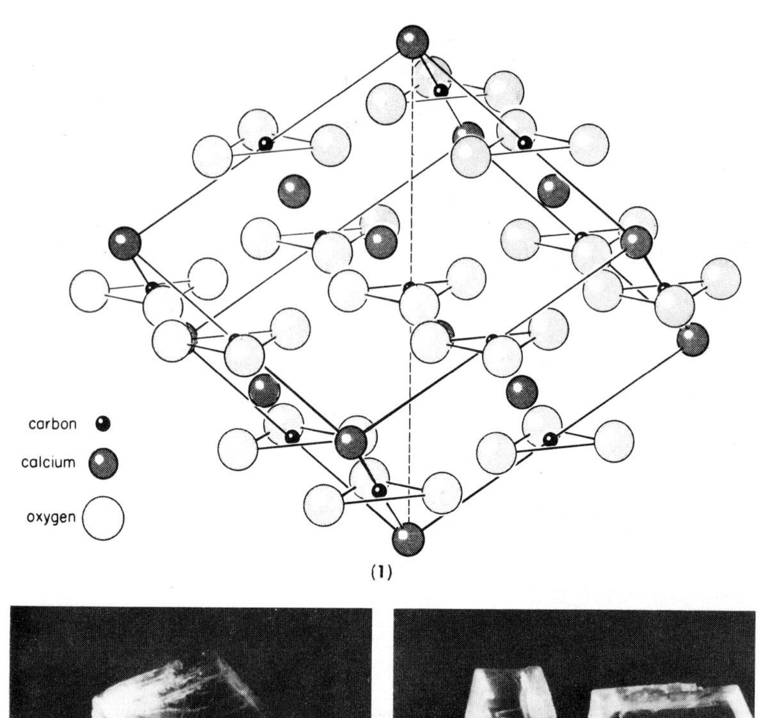

(1)

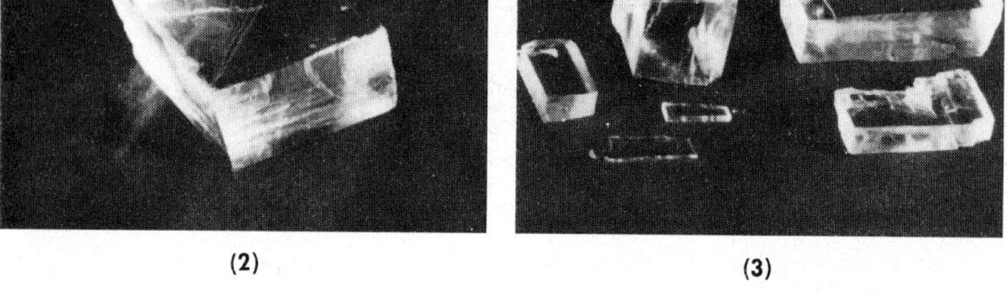

(2) (3)

PLATE VI (1) Model of the structure of calcite magnified approximately 10^8 times. (2) Cleavage fragment of calcite, approximately natural size. (3) Variously proportioned cleavage fragments of calcite, approximately natural size.

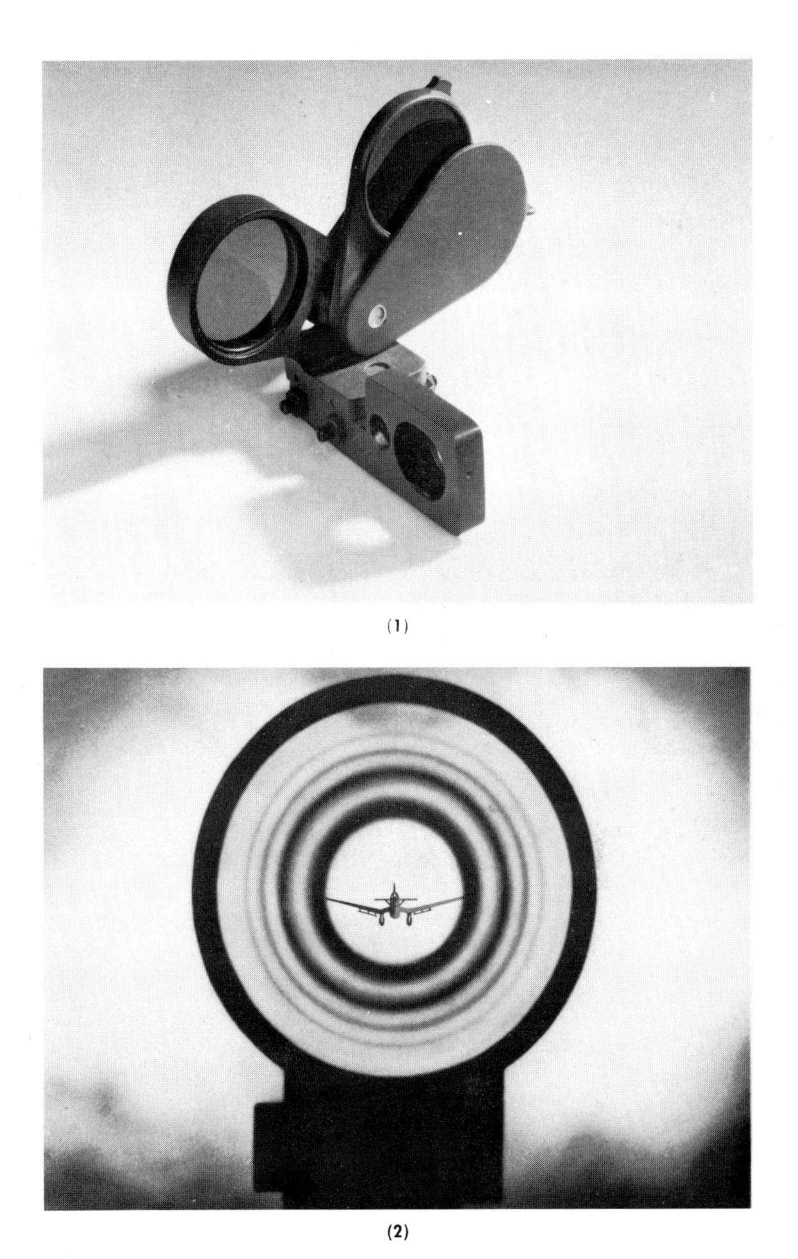

(1)

(2)

PLATE VII (1) The optical ring sight. (2) View through the optical ring sight, on target. (Courtesy of Polaroid Corporation).

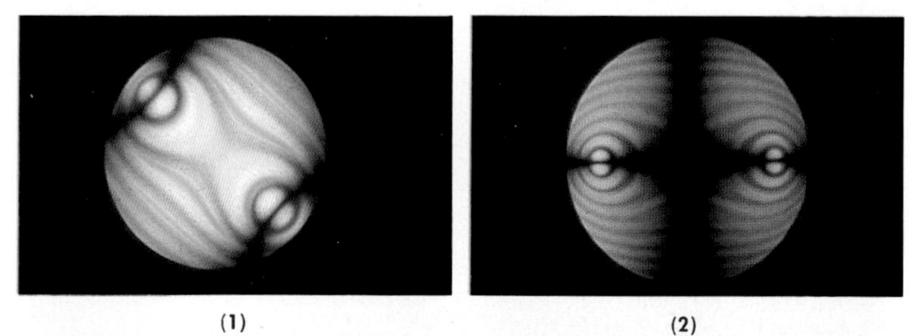

(1) (2)

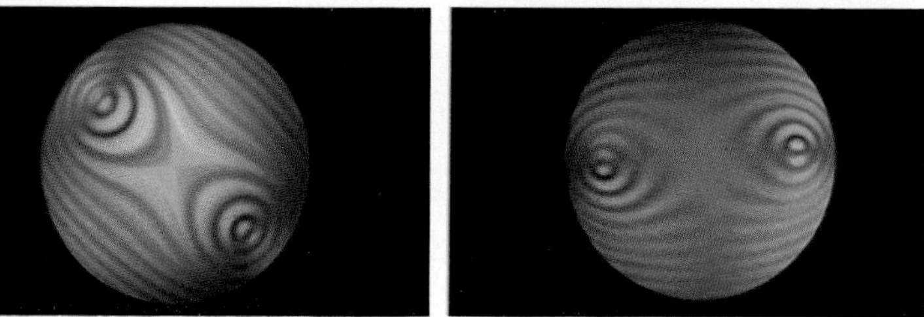

(3) (4)

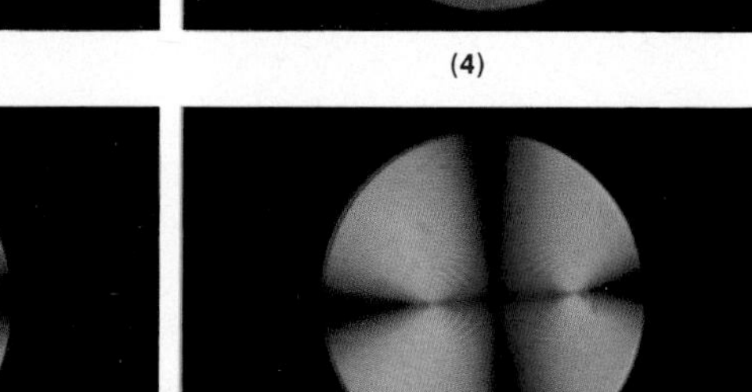

(5) (6)

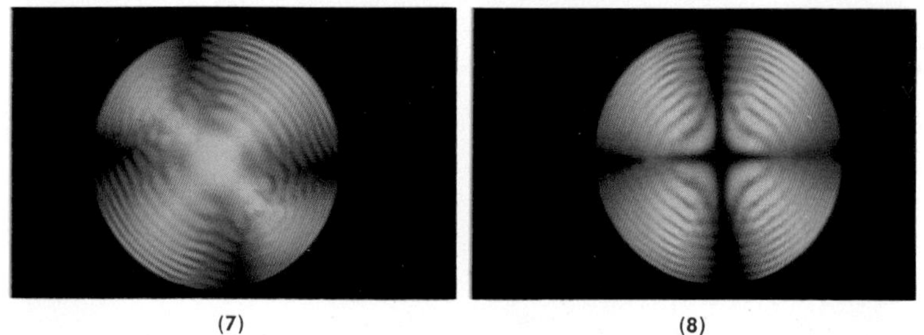

(7) (8)

PLATE VIII Interference figures (Photomicrographs courtesy of ERNST LEITZ *GMBH*, WETZLAR). (1) Muscovite, 45° position. (2) Muscovite, parallel position. (3) Muscovite, 45° position, with gypsum plate. (4) Muscovite, parallel position, with gypsum plate. (5) Titanite, 45° position. (6) Titanite, parallel position. (7) Brookite, 45° position. (8) Brookite, parallel position.

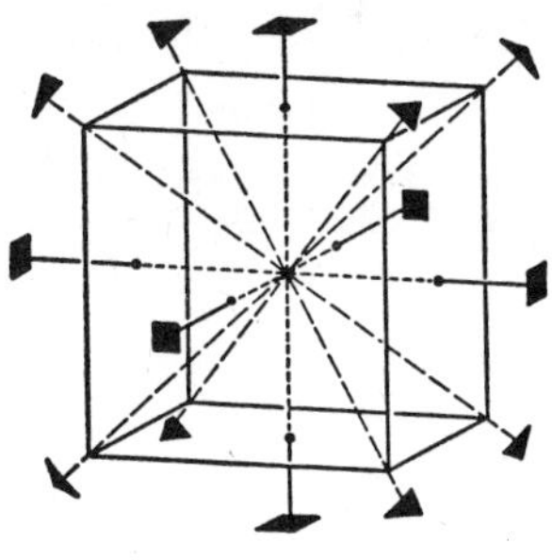

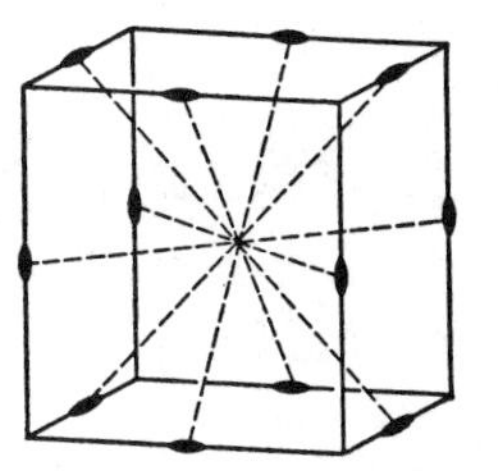

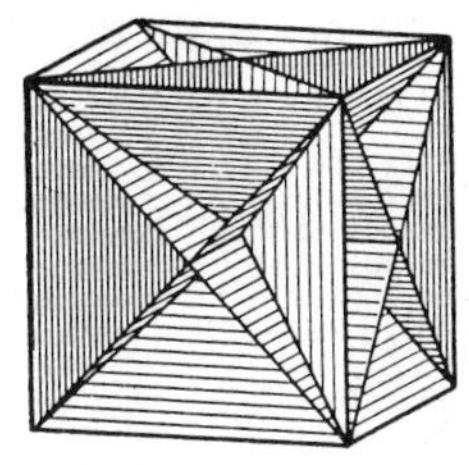

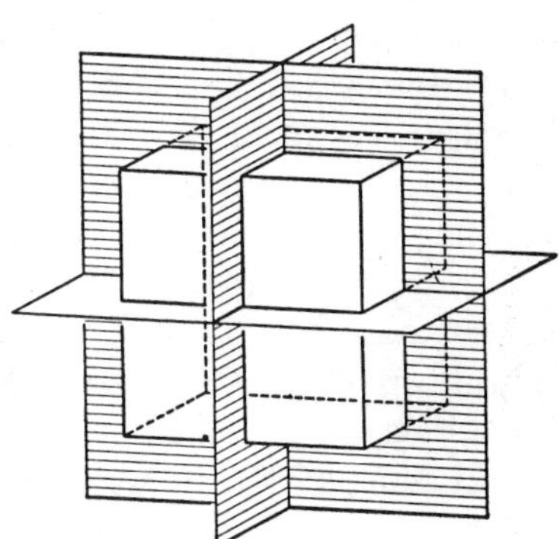

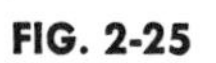

FIG. 2-25

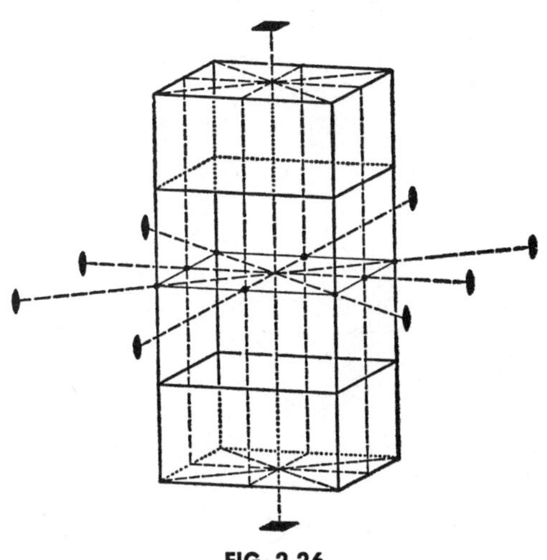

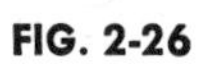

FIG. 2-26

3 *Directions and Planes, Miller Indices*

In any geometrical situation it is convenient to have a frame of reference to help us describe relative orientations of things. Latitude and longitude lines serve this function on a map: on graph paper we have the x and y axes, as in Fig. 3-1. In this figure two directions are shown by arrows starting at the origin, O and terminating at M and N respectively, and we could satisfactorily describe these directions by giving the coordinates of the first whole-numbered point (x,y) through which each passes. For the direction OM this would be 2,1. For the direction ON it would be 1,1.

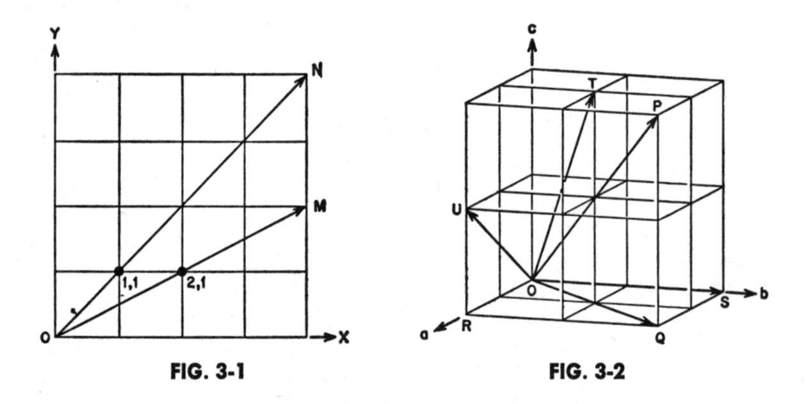

FIG. 3-1

FIG. 3-2

In a crystal, as we have seen in Chapter 2, we already have a convenient three-dimensional graph paper. In Fig. 3-2, for example, we have three mutually perpendicular directions, the a, b, and c axes, with repeat distances along them that are the same in all three directions—i.e., we have the axes of a cubic space lattice made up of unit cells like that in Fig. 2-10a. Such a lattice is shown in Fig. 3-2 with some directions indicated. We can now conveniently describe these directions, just as we did the directions in Fig. 3-1, but listing now three point coordinates in order of reference axes, *a*, *b*, and c.

OP	[111]	*OS*	[010]
OQ	[110]	*OT*	[112]
OR	[100]	*OU*	[201]

The *square brackets* are the crystallographer's convention for indicating a *direction*. Commas are unnecessary, but the digits are named in succession: "one one one," not "a hundred and eleven." Just as (x,y) refers to a point on the graph of Fig. 3-1 whose coordinates are not specifically given, so $[uvw]$ refers to a direction which is unspecified.

We can also use this lattice for describing the orientation of lattice planes by giving their intercepts on the three axes. For example, in Fig. 3-3 the intersection of the triangular planes

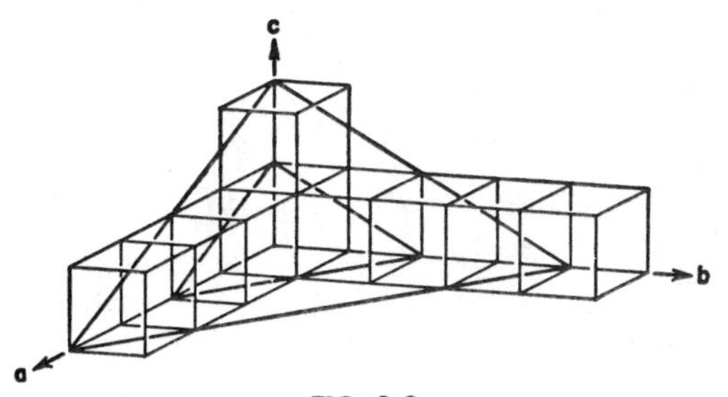

FIG. 3-3

with the *a* and *b* axes is twice as many units from the origin as their intersection with the *c* axis. Their intercepts are therefore 2,2,1. For reasons that will soon become clear, it is useful to describe the orientation of a plane by the reciprocals of its intercepts (in this case $\frac{1}{2}$, $\frac{1}{2}$, 1) rather than by its intercepts. Since the usefulness of this notation was first shown by Miller, such numbers, when appropriately converted to whole numbers, are called Miller indices.* To get whole numbers, in the case in Fig. 3-3,

* This system of indexing crystal faces was suggested by the Reverend Dr. W. Whewell, who was a professor of mineralogy at Cambridge University in

we multiply the reciprocals by 2 and get 112, called "one one two." Multiplying all intercepts, or reciprocals of intercepts, by the same number will of course not change the orientation of the plane, which is what we are interested in.

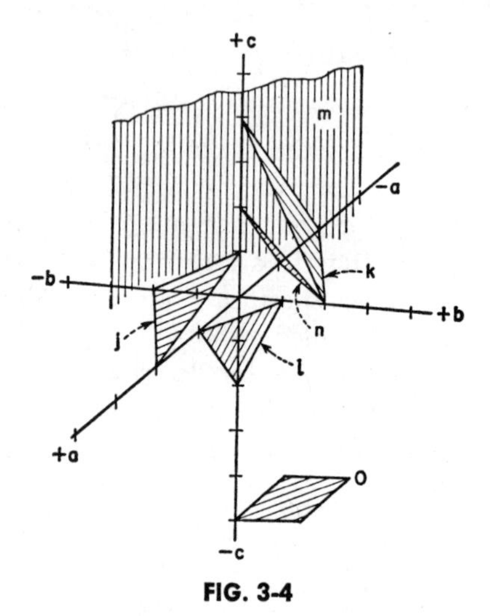

FIG. 3-4

Fig. 3-4 shows the axes of a lattice with a number of planes, each designated by a letter. It would be good practice for the reader to try to determine their Miller indices before referring to the list on the next page.

The Miller index for an intercept on the negative part of any axis is distinguished by a minus sign directly over it, e.g., $\bar{2}$. When a plane is parallel to one of the coordinate axes, we say it meets that axis at infinity, ∞. Accepting the convention that

---

England, although it had been used earlier in some of the work of C. S. Weiss, according to F. C. Phillips (*An Introduction to Crystallography*, Longmans, Green, 1956, p. 205). His student, William H. Miller, who assumed Dr. Whewell's professorship when the latter resigned in 1832, adopted this system and showed its usefulness in his *Treatise on Crystallography* in 1839. From this work such indices are known as Miller indices.

Haüy's law (Chapter 2) is now generally known as "the Law of Rational Indices." The word "rational" has the mathematical significance, referring to a number that can be expressed as a ratio of two whole numbers (unlike π, for example).

$1/\infty = 0$, the reciprocal of infinity thus gives us zero as the Miller index for that "intercept." The Miller indices of the planes in Fig. 3-4 are as follows:

j $(1\bar{1}2)$	l $(22\bar{1})$	n $(\bar{2}11)$
k $(\bar{2}21)$	m $(\bar{1}\bar{1}0)$	o $(00\bar{1})$

The *parentheses* are the crystallographer's convention for indicating a *plane*. Note that, since orientation, not position, is what is being described, we divide the Miller indices by their common factor so that plane o, which is shown in the figure with intercepts ∞, ∞, -5, is given the indices $(00\bar{1})$.

Just as $[uvw]$ refers to a direction which is unspecified, so (hkl) refers to a plane which is unspecified.

We can now label the cube faces of a crystal with the appropriate Miller indices, hkl, as shown in Fig. 3-5. The faces of many

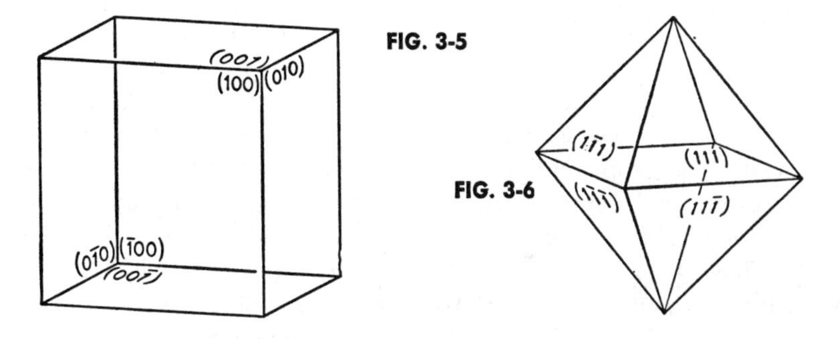

FIG. 3-5

FIG. 3-6

crystals with cubic unit cells show the eight-sided crystal form, octahedron, sketched in Fig. 3-6. Four of the eight sides are labeled.

It would be good practice for the reader to duplicate the sketch, adding the a, b, and c axes, and supply the Miller indices for the four faces on the far side. Compare your answer with that given at the end of this chapter.

In the hexagonal system there are four *crystallographic axes,** the c axis and three a axes in the plane normal to c, which are customarily labeled a_1, a_2, and a_3 as in Fig. 3-7, where c is normal to the paper. These a axes are shown in perspective in Fig. 3-8. All three have the same unit lengths. Only two of these, a_1 and

* A term used to designate the coordinate axes of a crystal.

a_2, were given in Fig. 2-10d, where they were labeled a and b. Indeed, as we shall see, the third one is, in some ways, unnecessary.

In Fig. 3-7 consider a plane normal to the paper (parallel to

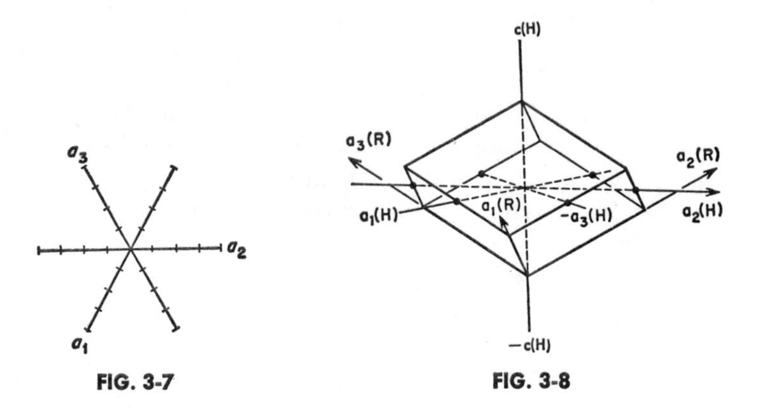

FIG. 3-7 FIG. 3-8

c), cutting the positive ends of a_1 and a_2 at 4 units, i.e., at the ends of the axis segments shown in the figure. If you lay a straight edge on the figure, you will see that this plane cuts the negative end of the third axis, a_3, at 2 units out. Its intercepts are thus 4, 4, −2, ∞. Taking reciprocals, to get Miller indices, we have $\frac{1}{4}$, $\frac{1}{4}$, $-\frac{1}{2}$, 0 and multiplying by four to get whole numbers, we have $(11\bar{2}0)$ as the indices of the plane. The general symbol is $(hkil)$. If the crystal has hexagonal symmetry, there will be five other planes like the $(11\bar{2}0)$ plane, on around the c axis. Naming them in counterclockwise order, they are $(\bar{1}2\bar{1}0)$, $(\bar{2}110)$, $(\bar{1}\bar{1}20)$, $(1\bar{2}10)$, and $(2\bar{1}\bar{1}0)$. Notice that the third index in all these planes is equal to the sum of the first two multiplied by −1—i.e., $i = -(h + k)$. This is always so, as you can find by testing other orientations of planes on Fig. 3-7. Therefore in writing hexagonal indices the third index is sometimes replaced by a centered dot, thus: $(11\cdot0)$. The intercept on the c axis is handled just as it is for other systems of axes.

This adaptation of the Miller indices to the hexagonal system is due to Auguste Bravais, and these indices are therefore known as Miller-Bravais indices.

For a discussion of the notation for directions in the hexagonal system, see *Structure of Metals,* by C. S. Barrett, and *Crystal Orientation Manual,* by E. A. Wood.

The use of the crystallographic axes with units and angles appropriate to the particular crystal you are trying to describe makes the task of description simpler, rather than more complicated as it might at first appear. In the case of the topaz crystal in Plate III and Figs. 2-13 through 2-15, it is not only simpler but more meaningful to use a notation for the faces that will indicate the small whole-number or building-block relationship between them than it would be to give their axial intercepts in terms of the same units (as, for example, fractions of a millimeter) in all directions.

The description of the unit-cell dimensions given in Fig. 2-10 is thus the description of the crystallographic axes appropriate for each crystal system. A crystal with a unit-cell shape belonging to one system cannot, in general, be simply and easily described with reference to crystallographic axes belonging to another system. The one outstanding exception is that crystals of the rhombohedral system are very conveniently described by using hexagonal crystallographic axes, as shown in Fig. 3-8. If you know the Miller indices of planes on one set of axes, you can find them for the other set of axes by the following formulas.

$$\begin{aligned} h_H &= h_R - k_R \\ k_H &= k_R - l_R \\ i_H &= l_R - h_R \\ l_H &= h_R + k_R + l_R \end{aligned}$$

where H refers to hexagonal axes and R refers to rhombohedral axes. For example, consider the face $(111)_R$. Visualize its orientation and you will see what $(hkil)_H$ must be. Check your result by the formulas above. (The answer will be given at the end of this chapter.)

Sometimes it may be more convenient to describe hexagonal crystals using orthogonal axes. Examination of Fig. 2-9 will suggest that a rectangular repeat unit could be selected which would give an orthorhombic unit cell twice the size of the hexagonal cell for the same substance. Such a cell is called an orthohexagonal cell, and it is sometimes used when one wishes to describe the crystal on orthogonal axes. The formulas for transforming Miller indices from hexagonal axes (*H*) to orthohexagonal (*O*) are as follows:

$$h_O = k_H + 2h_H$$
$$k_O = k_H$$
$$l_O = l_H$$

The crystallographic axes also provide a satisfactory coordinate system for describing the positions of atoms within the unit cell. To take the familiar cesium chloride example, in Fig. 2-1, the cesium atom is at 0, 0, 0, the origin of the coordinate system. The chlorine atom is at $\frac{1}{2}, \frac{1}{2}, \frac{1}{2}$; its coordinates are one-half unit along each of the three axes. In Fig. 2-2, the unit cell is so chosen that the chlorine atom is at 0, 0, 0 and the cesium atom is at $\frac{1}{2}, \frac{1}{2}, \frac{1}{2}$, but of course the two describe the same *structure* (arrangement of atoms), with cesium and chlorine alternating along the [111] direction.

When the units along different crystallographic axes are different, we can still describe positions in terms of fractions of these units without knowing their actual length. For example, in Fig. 2-10c you do not need to know the dimensions of the cell to know that an atom at $\frac{1}{2}, \frac{1}{2}, \frac{1}{2}$ is at its center. An atom at $\frac{1}{2}, 0, \frac{1}{2}$ in this cell is shown in Fig. 3-9. Since its coordinate along the *b* axis is

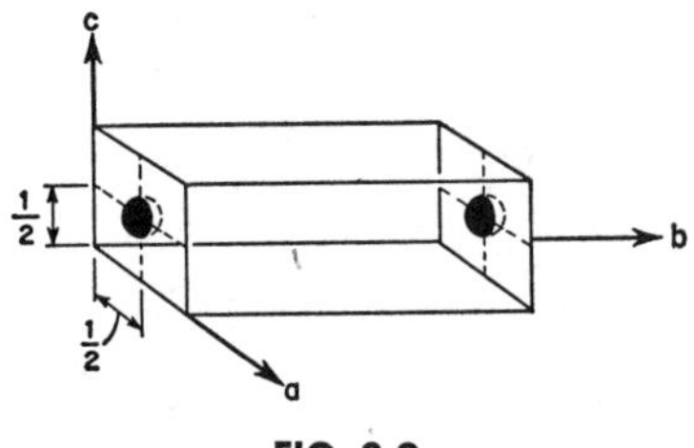

FIG. 3-9

zero, its center is at the boundary of the cell so that half the atom is in the cell and half out of the cell. The same atom is repeated at the far end of the cell, however, since here the *b* coordinate is zero for the neighboring cell. Here again half of the atom is in the cell shown in Fig. 3-9. So there is a total of two halves or one atom in the cell at the position $\frac{1}{2}$, 0, $\frac{1}{2}$, just as the 8 eighths of an atom at 0, 0, 0 in Fig. 2-2 gave one atom at position 0, 0, 0.

In the cubic system a plane with a given set of Miller indices is normal to the direction with the same indices, as shown in Fig. 3-10. Examination of this figure will show why this must be

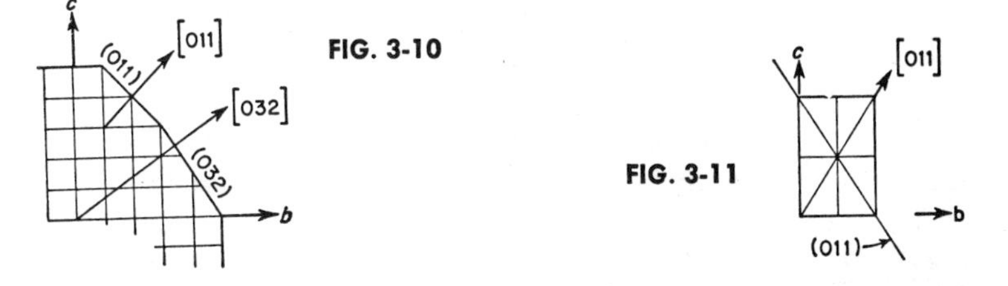

FIG. 3-10

FIG. 3-11

so for the ($0kl$) planes (parallel to the a axis which is normal to the page) and the [$0kl$] directions (normal to a and therefore parallel to the page) shown there. This relationship depends upon the orthogonality of the axes ($b \perp c$, i.e., $\alpha = 90°$) and the fact that the units along both axes are the same. When this is not so, a plane and direction that have the same indices will not be normal to each other. Fig. 3-11 shows the (011) plane and [011] direction for a tetragonal crystal ($b \neq c$).

Note on usage of "a," "b" and "c": These letters were first used in Fig. 2-4 to indicate the directions of what were later called "crystallographic axes." In Fig. 2-10 and the accompanying text, the same letters were used to refer to the lengths of units along these axes. Crystallographers use them in both ways, as we just have in the preceding paragraph. In earlier crystallographic literature "a_0, b_0, c_0" were commonly used when referring to the unit lengths, but more recent practice has been to omit the subscript zero.

When a crystal of sodium chloride grows from a water solution of the salt as the water evaporates, it grows with the six faces of a cube. The salt grains in table salt show the cubic form nicely under a magnifying glass. Various processes of preparation used by different companies result in different habits of growth. In one brand, the crystals occur as isolated cubes, all very nearly the same size. In another, the cubes stick together in clusters and are of various sizes. In Kosher salt, the crystals show skeletal growth, parts of the cube not being filled in solidly. The structure of sodium chloride is not the same as that of cesium chloride, but both arrangements have all the symmetry elements shown for a cube at the end of Chapter 2. Each of the cube faces on the sodium chloride crystal is therefore indistinguishable from any other of the cube faces since they are all equivalent, being related by symmetry.

The name for all those planes that are symmetrically equivalent is a *form*. When the symmetry of the crystal is known, one needs only to mention the indices of one plane of the form and the rest spring into existence because of the symmetry elements. When one plane is used to represent the whole form in this way, it is enclosed in curly brackets, thus: $\{hkl\}$. To represent the cube in Fig. 3-5, for example, we need only write {100}. If we then operate on the (100) plane with all the symmetry operations indicated for the cube at the end of Chapter 2, we find that we generate [in addition to (100)] (010), (001), $(\bar{1}00)$, $(0\bar{1}0)$ and $(00\bar{1})$. Similarly, angular brackets, $\langle uvw \rangle$, refer to all those *directions* that are symmetrically equivalent.

Suppose we had a crystal which belonged to the orthorhombic system (Fig. 2-10c). It might have faces like those on the brick in Fig. 1-7 with the symmetry shown in Figs. 1-8 and 1-9.

Let the three 2-fold symmetry axes of the object be its *a, b,* and *c* axes in the usual orientation. Since the symbol {100} refers to all those faces generated by the symmetry operations once the (100) face has been set up, we set up the (100) face and operate on it with the 2-fold axes (Fig. 3-12).

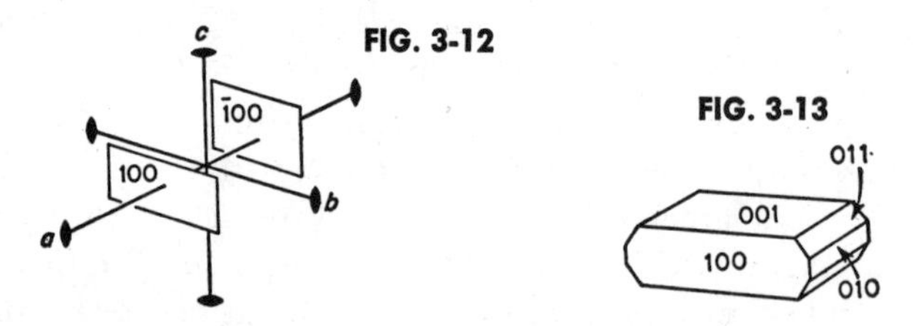

FIG. 3-12

FIG. 3-13

This is a form with only two faces, an "open form." Clearly, when an orthorhombic crystal grows with {100} faces, there will also be other faces on the crystal. Fig. 3-13 is a drawing of a crystal of barite ($BaSO_4$) showing the forms {100}, {010}, {001} and {011}. We have seen in Fig. 3-12 that the form {100}, in a crystal with this symmetry, consists of (100) and $(\bar{1}00)$. Similarly {010} consists of (010) and $(0\bar{1}0)$, and {001} consists of (001) and $(00\bar{1})$, but how about {011}? The face (011) is parallel to *a*, but cuts *b* and *c* at unity (not necessarily the same distance from the origin in this system). Act on it with the 2-fold axis that coincides with *c* and you get $(0\bar{1}1)$. Now if you act on this roof-like pair of faces with either the 2-fold axis along *a* or that along *b*, you will get $(0\bar{1}\bar{1})$ and $(01\bar{1})$—i.e., two faces parallel to *a* and cutting the negative end of the *c* axis at unity. So this form has four faces.

All four of the {011} faces are parallel to *a*. All those faces that are parallel to a given direction are called a *zone*, and the direction to which they are parallel is the axis of the zone, or the *zone axis*. In this case it is *a*—i.e., [100].

Given two of the faces of a zone (e.g., $(0\bar{1}1)$ and (011)), you can

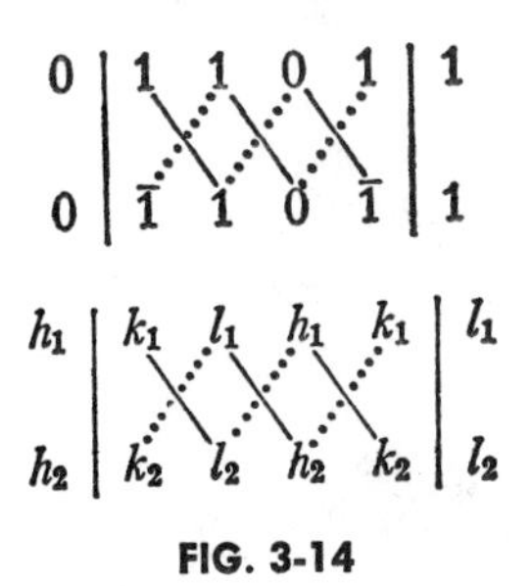

FIG. 3-14

find the zone axis by writing them as shown in Fig. 3-14 and then subtracting the dotted-line products from the solid-line products, thus:

$$1 \times 1 - 1 \times (-1) = 2$$
$$1 \times 0 - 1 \times 0 = 0$$
$$\bar{1} \times 0 - 1 \times 0 = 0$$

Since factoring does not change the direction given by the Miller indices, we may divide through by 2 and get [100].

The zonal equations we have just used may be written: $k_1l_2 - k_2l_1 = u$; $l_1h_2 - l_2h_1 = v$; $h_1k_2 - h_2k_1 = w$. If any face (hkl) lies in a zone $[uvw]$, $hu + kv + lw = 0$.

The zonal equations, in a different form, can also be used to find the indices of a face if it is known to belong to two given zones. To go back to Fig. 3-13, what face belongs both to zone [100] and to zone [001]?

For the answer we set up the equation thus:

$$v_1w_2 - v_2w_1 = h; \quad w_1u_2 - w_2u_1 = k$$
$$u_1v_2 - u_2v_1 = l$$

In this case:

$$\begin{array}{c|cccc|c} 1 & 0 & 0 & 1 & 0 & 0 \\ 0 & 0 & 1 & 0 & 0 & 1 \end{array}$$

$0 - 0 = 0$; $1 - 0 = 1$; $0 - 0 = 0$. The answer is (010).

The use of Miller indices has simplified the calculation of zonal relations and also of the angle between any pair of faces (i.e., any pair of atomic planes) in any system. The formulas for the latter may be found in *Structure of Metals,* by C. S. Barrett.

* * *

Answer to question in Chapter 3 Miller indices of the faces of an octahedron: Fig. 3-15 shows the octahedron with back faces indexed. The

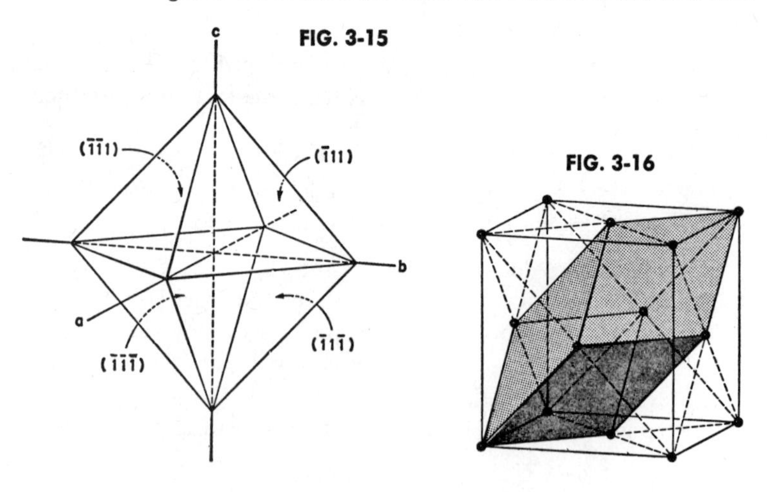

FIG. 3-15

FIG. 3-16

symmetry of the octahedron is the same as that of the cube, as given in Fig. 2-26. All the directions normal to the octahedral faces are 3-fold axes. All the ⟨100⟩ directions are 4-fold axes.

* * *

Answer to question in Chapter 3 concerning hexagonal indices of the face $(111)_R$: Since this meets the three symmetrically placed rhombohedral axes all at the same distance from the origin (bottom point of the rhombohedron in Fig. 3-8), it must be normal to the *c* axis and therefore have the hexagonal indices (0001). Calculation from the formulas gives (0003), which should always be factored to give the smallest whole-number indices, i.e., (0001).

* * *

Problem: In the unit cell of sodium chloride, the sodium atoms are in the positions 000, $\frac{1}{2}\frac{1}{2}0$, $\frac{1}{2}0\frac{1}{2}$ and $0\frac{1}{2}\frac{1}{2}$. The chlorine atoms are in the positions $00\frac{1}{2}$, $0\frac{1}{2}0$, $\frac{1}{2}00$ and $\frac{1}{2}\frac{1}{2}\frac{1}{2}$. (1) Sketch the positions of the atoms. (2) On a (100) face of sodium chloride, are the atoms all of the same kind, or does this atomic plane contain both kinds of atoms? (3) Answer the same question for the (111) plane. (4) Answer the same question for the (110) plane. (5) In sodium chloride $a = 5.627$ Å. How many atoms are there per cubic centimeter? (Answers are given at the end of Chapter 4.)

* * *

Answers to questions at the end of Chapter 2

1. Fig. 3-16 shows the lattice points of a face-centered cubic cell with a primitive cell shaded in.
2. The primitive cell is rhombohedral.
3. Since the f.c.c. cell contains four lattice points, it is four times the size of the primitive cell.

4 *The Three-Dimensional Crystal on Two-Dimensional Paper*

Crystals are three-dimensional objects. They should be held in the hand and examined from all sides. Even with such examination it is not always possible to determine what the symmetry of the crystal lattice is because of the chance effects of growth conditions. If a cubic crystal growing from solution happens to have more atoms added to it along one direction than another, it may grow long in that direction (Fig. 4-1). Its outside shape will no longer be that of a cube, because of the unequal rates of growth, but of course its unit cells will still be cube-shaped as shown, for example, by x-ray diffraction; it will still be a cubic crystal. As can be seen from Fig. 4-1, the interfacial angles are in no way

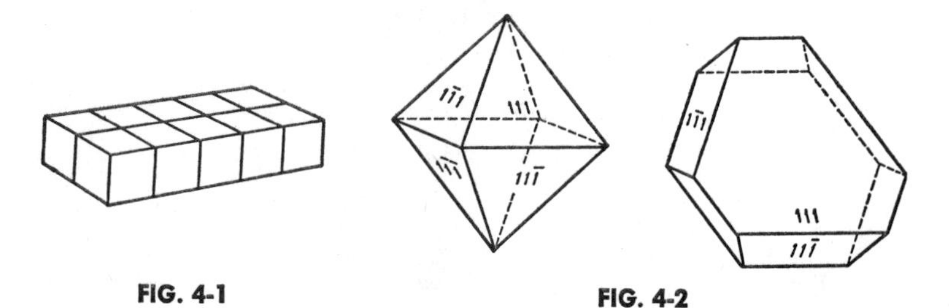

FIG. 4-1 **FIG. 4-2**

changed by such distortion: consideration of the way in which the crystal is built up convinces us that adding more material at one end cannot change the angles between the planes. Fig. 4-2 shows the octahedron, {111}, with equal and unequal development of its eight faces. Fig. 4-3 shows three dodecahedrons, the twelve-sided (*do,* 2, + *deca,* 10) form with {110} faces. Clearly, then, the internal symmetry of the crystal, its structural symmetry, is revealed by the *orientation* of its faces, not their size or shape, and it is useful to have a way of showing clearly on our two-dimensional piece of paper what the orientations of the faces on a crystal are without regard to their size or shape.

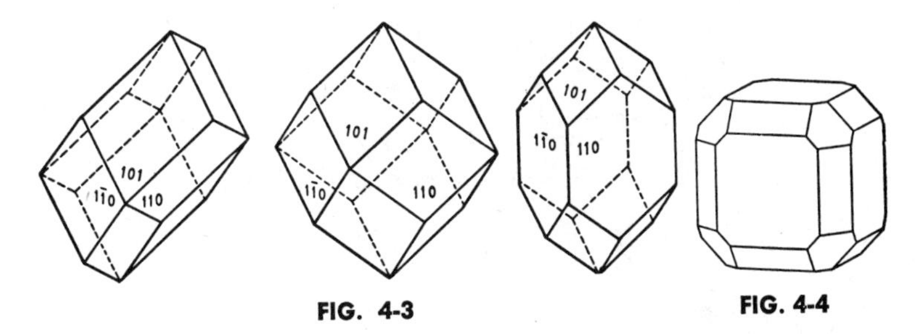

FIG. 4-3

FIG. 4-4

Several methods can be used. Only one will be described in this chapter. Fig. 4-4 is a sketch of a cubic crystal on which the faces of three different forms developed as it grew. It is viewed in the conventional way, from a little above and to the right, as described in Tutton's instructions for drawing crystals.* From this view we see 3 of the 6 cube faces, {100}, 4 of the 8 faces of the octahedron, {111}, and 6 of the 12 faces of the dodecahedron, {110}.

If we could put this crystal in the middle of a sphere and draw an imaginary line from the center point of the sphere through each face, perpendicular to the face, extending these until each one met the sphere at a point, then this group of points would be a *spherical projection* of the faces of the crystal (Fig. 4-5). If we now take away the crystal, the arrangement of points on the sphere reveals to us the symmetry of the crystal without giving any information about its shape or size. (Examine the spherical projection and find each kind of symmetry axis: 4, 3, 2.)

But the sphere is a bit unhandy to carry around, so this essential information must be transferred in some orderly way to a two-dimensional plane. One way would be to put a plane at the

* A. E. H. Tutton, *Crystallography and Practical Crystal Measurement.* Macmillan, 1922. Chapter XXV: Drawing Crystals.

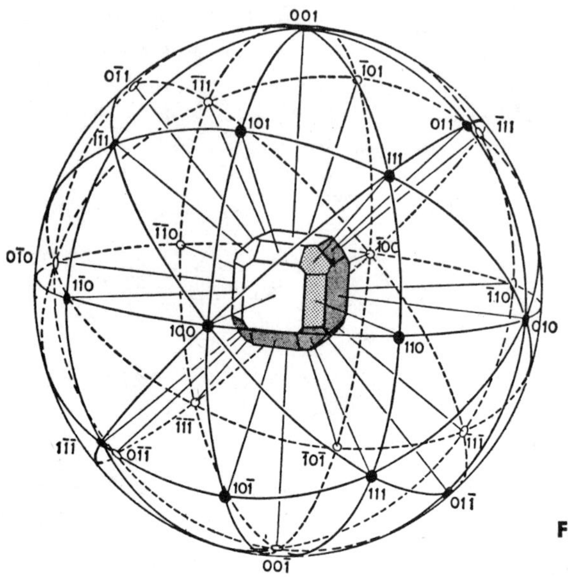
001
$0\bar{1}1$
$\bar{1}01$
$\bar{1}\bar{1}1$
101
011
$\bar{1}11$
$1\bar{1}1$
111
$\bar{1}\bar{1}0$
$0\bar{1}0$
$\bar{1}00$
$1\bar{1}0$
$\bar{1}10$
100
110
010
$\bar{1}\bar{1}\bar{1}$
$1\bar{1}\bar{1}$
$0\bar{1}\bar{1}$
$\bar{1}0\bar{1}$
$\bar{1}1\bar{1}$
$10\bar{1}$
$11\bar{1}$
$01\bar{1}$
$00\bar{1}$

FIG. 4-5

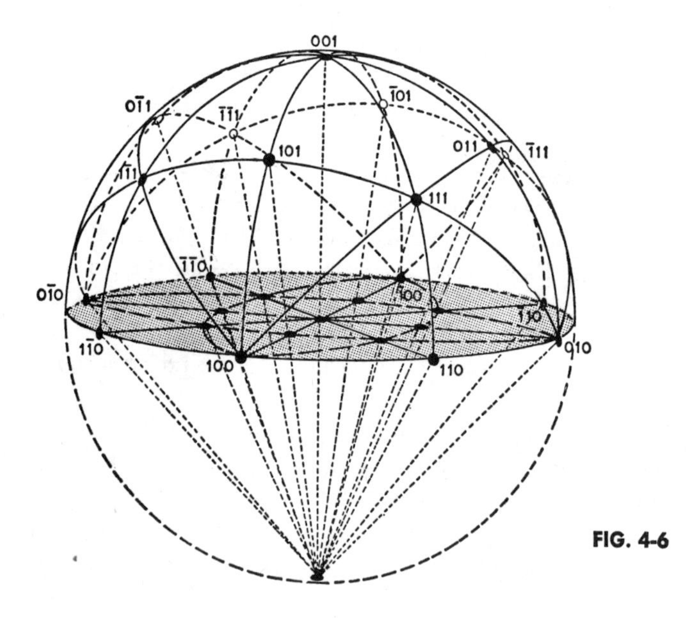
001
$0\bar{1}1$
$\bar{1}01$
$\bar{1}\bar{1}1$
101
011
$\bar{1}11$
$1\bar{1}1$
111
$\bar{1}\bar{1}0$
$0\bar{1}0$
$\bar{1}00$
$\bar{1}10$
$1\bar{1}0$
100
110
010

FIG. 4-6

top of the sphere, normal to the [001] direction, and extend all the normals until each one met this plane in a point. These points $(hkl)_p$ would constitute a "gnomonic projection" of the planes. This projection is sometimes used, but it has the disadvantage that many points will lie beyond the edge of any plane of reasonable size.

An alternative which we will describe is the *stereographic projection* (*stereo:* 3-dimensional, solid, + *graphic,* having to do with writing or drawing). Connect each point in the upper half of the spherical projection of Fig. 4-5 with the "South Pole" of the sphere. These connecting lines pierce the equatorial plane in points which constitute the stereographic projection (Fig. 4-6). Every plane in the upper half of the crystal is represented by a spot on the projection. Even the $\{hk0\}$ faces, which were lost in the gnomonic projection, are preserved here, since their spherical-projection points are the same as their stereographic-projection points.

Looking down on the projection (Fig. 4-7), we see how well it

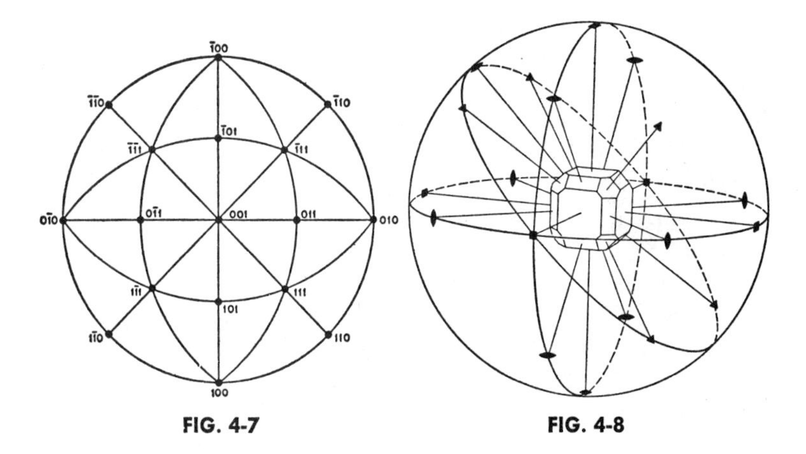

FIG. 4-7

FIG. 4-8

portrays the symmetry of the crystal. By turning the diagram in Fig. 4-6 upside down, we could project the faces of the lower half of the crystal onto the same plane. Points representing these faces are customarily indicated by circles rather than dots. The full symmetry of forms $\{100\}$, $\{110\}$, and $\{111\}$, shown in Fig.

4-4, is the same as that for the cube, shown in Fig. 2-25. Some of these symmetry elements are shown in spherical projection in Fig. 4-8, but clearly the diagram would get too complicated if we tried to show all of them. If we transfer these to the stereographic projection, however, the rotation axes (which met the sphere in points) will occur as points, and the mirror planes (which met the sphere in great circles as shown in Fig. 4-6) will occur as segments of circles (Fig. 4-9).

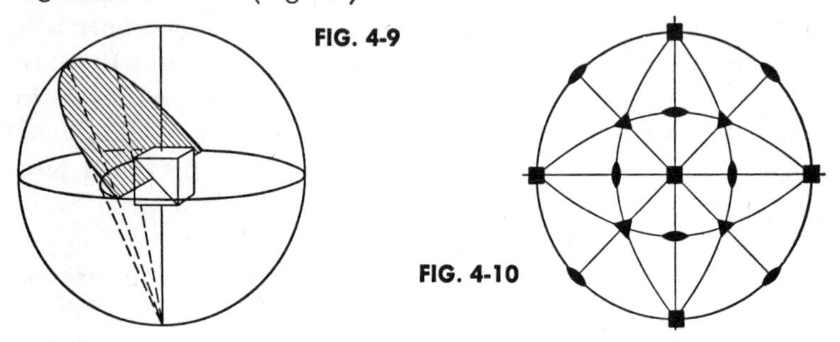

FIG. 4-9

FIG. 4-10

The completed stereographic projection of all the symmetry elements of Fig. 2-25 is given by Fig. 4-10. It is as though we were looking down on the vertical 4-fold axis from the top, seeing the other two (horizontal) 4-fold axes, each with its two ends marked by black squares on the rim of the projection. With this as a start, identify the remaining features of Fig. 4-11 with those of Fig. 2-25.

This representation of symmetry elements was used in Fig. 1-9, which is, in fact, the stereographic projection of the symmetry elements of the brick.

From Fig. 4-7 we can see that the position of any point on the stereographic projection is precisely determined by the orientation of the normal to the plane, which, in turn, is determined by the plane that it represents. Good directions for plotting points on a stereographic projection with the aid of a stereographic "net" or grid are given in the references at the end of this book. Knowing how the projection is produced (Fig. 4-6), we could plot points without the aid of such a net, although it takes less time with the net. However, there are many powerful uses we can make of the projection without plotting points precisely.

In Chapter 3 we found that {100} in the cubic system is a form with six faces, whereas in the tetragonal system it has only four faces. We can show this very neatly with the stereographic projection.

In Fig. 4-11 the upper figures are the projection of the symmetry elements of the most symmetrical point group of the cubic and tetragonal systems, respectively. Underneath each is the projection of the {100} faces. The little circle in the left figure represents the $(00\bar{1})$ face which would be on the far side, away from us, since we are always looking at the positive end of the c axis in these projections.

In choosing the {100} form we are choosing a special form. The normals to its faces coincide with symmetry axes. We could see the effect of the symmetry elements better if we chose a "general" form: one whose faces were not normal to any symmetry axis or plane. Suppose we choose an $\{hkl\}$ with no zero values and no two values alike, for example, {123}. Fig. 4-12

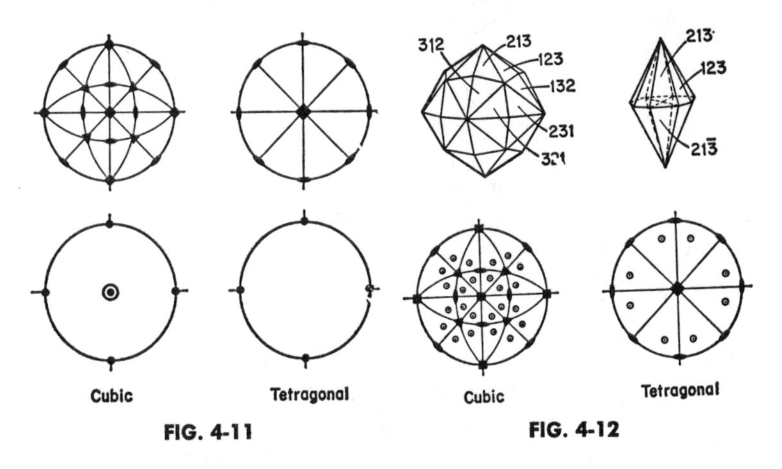

FIG. 4-11 FIG. 4-12

shows the form {123} for the two point groups given in Fig. 4-11. Below each, the stereographic projection of the faces of this general form is shown combined with the stereographic projection of the symmetry elements.

There are many interesting things to notice here. In the cubic case, since all three axes are symmetrically equivalent, the indices are permuted among all three. In other words, if the face (123) exists, then, because of the symmetry elements of the point group, the following faces also exist: (312), (231), (213), (321), (132); $(\bar{1}23)$, $(3\bar{1}2)$, $(23\bar{1})$, $(2\bar{1}3)$, $(32\bar{1})$, $(\bar{1}32)$; $(1\bar{2}3)$, $(31\bar{2})$, $(\bar{2}31)$, $(\bar{2}13)$, $(3\bar{2}1)$, $(13\bar{2})$; $(12\bar{3})$, $(\bar{3}12)$, $(2\bar{3}1)$, $(21\bar{3})$, $(\bar{3}21)$, $(1\bar{3}2)$; $(\bar{1}\bar{2}3)$, $(3\bar{1}\bar{2})$, $(\bar{2}3\bar{1})$, $(\bar{2}\bar{1}3)$, $(3\bar{2}\bar{1})$, $(\bar{1}3\bar{2})$; $(\bar{1}2\bar{3})$, $(\bar{3}\bar{1}2)$, $(2\bar{3}\bar{1})$, $(2\bar{1}\bar{3})$, $(\bar{3}2\bar{1})$, $(\bar{1}\bar{3}2)$; $(1\bar{2}\bar{3})$, $(\bar{3}\bar{1}2)$, $(2\bar{3}\bar{1})$, $(2\bar{1}\bar{3})$, $(\bar{3}2\bar{1})$, $(\bar{1}\bar{3}2)$; $(1\bar{2}\bar{3})$, $(\bar{3}1\bar{2})$, $(\bar{2}\bar{3}1)$, $(\bar{2}1\bar{3})$, $(\bar{3}\bar{2}1)$, $(1\bar{3}\bar{2})$. There are 48 faces all together, and the form is called a hexoctahedron (6 × 8). It is found commonly on garnet crystals, usually in combination with the dodecahedron, {110}, as in Fig. 4-13.

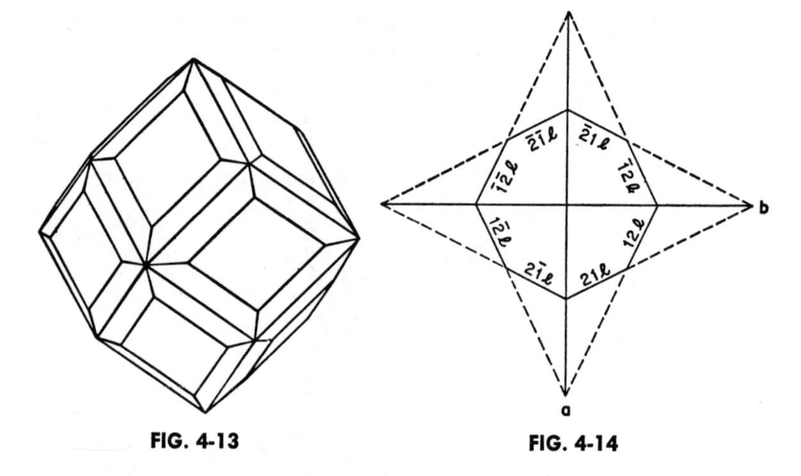

FIG. 4-13

FIG. 4-14

In Fig. 4-12, note the relation of the {123} points (the stereographic projection of the general form) to each of the symmetry elements. For example, because of the symmetry plane parallel to the paper, there is a circle and a dot at each position: the circle from the spherical projection point on the far side of the paper, and the dot from the spherical projection point on the near side of the paper. Note the 4-fold symmetry around the 4-fold axis and the mirror-symmetry relation between the points on opposite sides of the mirror planes.

The tetragonal case is simpler. There are fewer symmetry elements, and the general form has only 16 faces. Here only the *a* and *b* axes (sometimes called a_1 and a_2 axes) have the same unit lengths and are therefore permutable. The form has the following faces: (123), (213); $(1\bar{2}3)$, $(\bar{2}13)$; $(\bar{1}23)$, $(2\bar{1}3)$; $(\bar{1}\bar{2}3)$, $(\bar{2}\bar{1}3)$; $(12\bar{3})$, $(21\bar{3})$; $(1\bar{2}\bar{3})$, $(\bar{2}1\bar{3})$; $(\bar{1}2\bar{3})$, $(2\bar{1}\bar{3})$; $(\bar{1}\bar{2}\bar{3})$, $(\bar{2}\bar{1}\bar{3})$. All of the planes meet the *c* axis at one-third its unit length. In the tetragonal system the *c* axis has a different unit length from the other two, the length depending upon the particular substance in question.

In Fig. 4-14, a section in the *ab* plane through the upper right diagram of Fig. 4-12, the third index is given as unspecified (l) since, from this diagram, the intercept on the *c* axis cannot be determined.

Referring back to Fig. 1-10, we recall that a center of symmetry relates any feature of the crystal to a similar feature on the opposite side of the center. For every dot in the stereographic projection of planes related by a center of symmetry, there will be a circle on the opposite side of the center of the projection, the same distance out from the center (Fig. 4-15). The dot and circle represent the projections of a pair of parallel faces on opposite sides of the crystal. In other words, they are the opposite ends of a line going through the center of the crystal. Note that each of the point groups shown in Fig. 4-12 has a center of symmetry.

Fig. 4-16 is a sketch of a crystal of the tetragonal mineral scheelite (calcium tungstate, $CaWO_4$) in which the $\{131\}$ form is shaded and two faces of $\{101\}$ are labeled. If we could turn the crystal around, we would see that the *c* axis is a 4-fold axis of

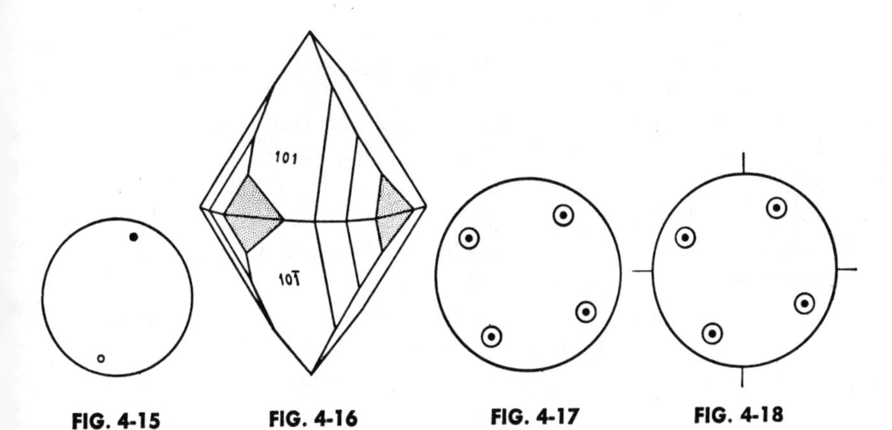

FIG. 4-15 FIG. 4-16 FIG. 4-17 FIG. 4-18

symmetry with a mirror plane normal to it. The four kite-shaped pairs of {131} faces follow, head to tail, like horses on a merry-go-round, around the equator of the crystal so that no other planes or axes of symmetry are present. Fig. 4-17 is the stereographic projection of these {131} planes. Given just that figure alone, it would appear that this form had four symmetry planes normal to the paper as well as four 2-fold axes lying in the plane of the paper. However, it is clear from looking at the crystal that this is not so. How can we avoid this difficulty? We must put the projection of the general form into its proper frame of reference: the crystal lattice. In Fig. 4-18 the positions of the crystallographic axes normal to *c* are marked at the edge of the projection. Now it is clear that the plotted form is skewed around with reference to the directions of the edges of the tetragonal unit cell (i.e., to the crystallographic axes) and that the only symmetry elements common to both are a 4-fold axis normal to the paper and a mirror plane parallel to the paper, plus the center of symmetry which this combination of symmetry elements implies.

The stereographic projection provides us with an easy way to discover what elements are implied when certain elements are given. In Fig. 4-19, for example, we have a 2-fold axis normal to

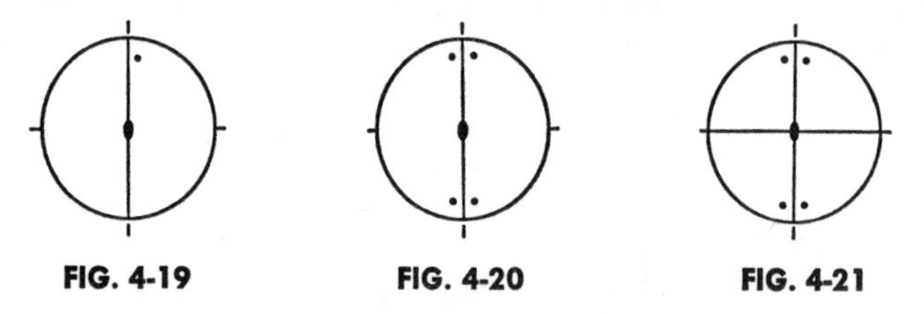

FIG. 4-19 **FIG. 4-20** **FIG. 4-21**

the paper and a mirror containing the 2-fold axis. One plane of a general form is plotted as a point. If now we act on this point with the two symmetry operations, the additional points shown in Fig. 4-20 are generated. Inspection of this figure shows us that the projected points are also related by a second mirror plane normal to the first and also normal to the paper, as shown in Fig. 4-21. If we had started with the two mirror planes, we would have discovered that the 2-fold axis came into existence. Try the same experiment with a 4-fold axis and a mirror plane parallel to it.

Then make up your own combinations. If you try to put two symmetry elements together that cannot belong in the same point group, you will soon find this out by the difficulties you get into with the projection of the crystal planes.

We can also now follow more easily the operations of an inversion axis, such as the 6-fold inversion axis shown in Fig. 4-22.

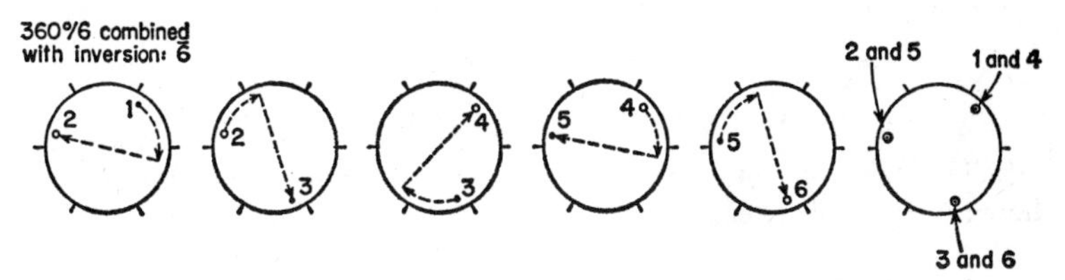

FIG. 4-22

This is a particularly interesting case since from the finished projection of the general form it is clear that we could have achieved the same result with a 3-fold rotation axis and a mirror plane normal to it. Note that a center of symmetry is the same as a one-fold inversion axis (Fig. 4-15).

* * *

Problem: In each of the stereographic projections below, the symmetry elements of some point group are given, together with one point of a general form. Act on this point with the symmetry elements and produce the remaining projected points of this general form. The crystallographic axes are indicated at the rim of each projection. They are not symmetry axes unless so indicated.

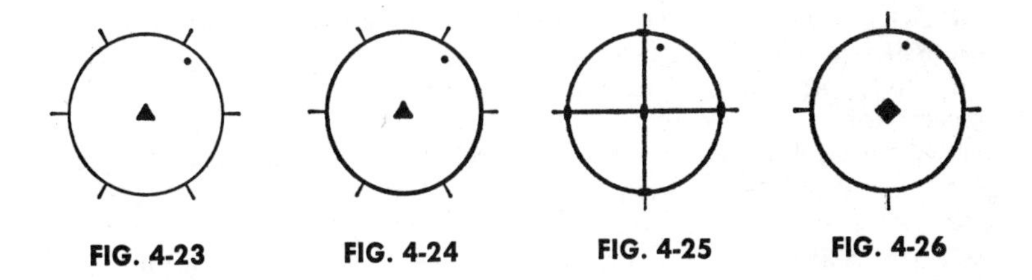

FIG. 4-23 FIG. 4-24 FIG. 4-25 FIG. 4-26

1. A 3-fold axis normal to the paper (Fig. 4-23).
2. A 3-fold axis normal to the paper and a mirror plane lying in the plane of the paper (Fig. 4-24).

3. Three mutually perpendicular mirror planes (two normal to the paper and one lying in the paper) with three 2-fold axes, each lying along the intersection of a pair of planes (Fig. 4-25). This is the familiar orthorhombic symmetry of the brick.

4. A 4-fold axis normal to the paper and a mirror plane lying in the paper (Fig. 4-26).

5. Which of these forms has a center of symmetry? (Answers are given at the end of Chapter 5.)

* * *

Answers to questions at the end of Chapter 3

1. The positions of atoms in the unit cell of sodium chloride are shown in Fig. 4-27. Because of the repetitious nature of the structure of

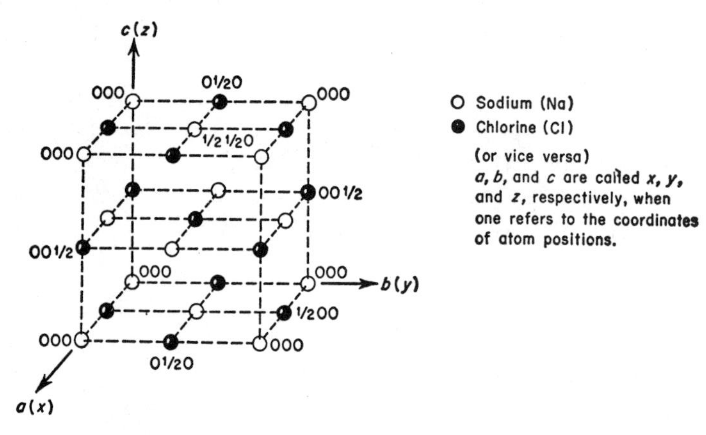

FIG. 4-27

a crystal, if the coordinate of an atom's position along the *a* axis is 1, this means it is one whole unit-cell length away from the origin and therefore at the beginning of the next unit cell, that is, at zero. Thus, all corner atoms of the cells have the coordinates 0, 0, 0. For the same reason, the atom at the center of the top face has the coordinates $\frac{1}{2}$, $\frac{1}{2}$, 0, just as the atom at the center of the bottom face does.

2. (100): Sodium and chlorine atoms alternate (Fig. 4-28).

3. (111): Each atomic plane of {111} orientation consists of one type of atom only. Sheets of sodium atoms alternate with sheets of chlorine atoms (Fig. 4-29).

4. (110): Each atomic plane of {110} orientation consists of rows of sodium atoms alternating with rows of chlorine atoms (Fig. 4-30).

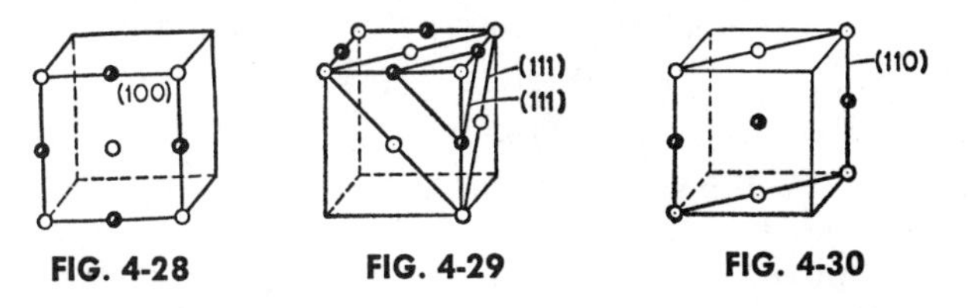

FIG. 4-28 FIG. 4-29 FIG. 4-30

5. In the cubic unit cell, 5.627 Å. on an edge, there are 8 atoms. (See Fig. 4-27: whole atom at the center, 1, plus $\frac{1}{8}$ atom at each of eight corners, 1, plus $\frac{1}{2}$ atom centered in each of six faces, 3, plus $\frac{1}{4}$ atom in the middle of each of twelve edges, 3, makes a total of $1 + 1 + 3 + 3 = 8$ atoms). Therefore in 1 cc there are $8 \times (1/5.627 \times 10^{-8})^3 = 4.5 \times 10^{22}$ atoms.

Compare this figure with that for cesium chloride at the end of Chapter 2. If you want to calculate the number of atoms per cc in other substances, you will find the unit cell dimensions in the *Handbook of Chemistry and Physics* published periodically by the Chemical Rubber Publishing Co., Cleveland, Ohio, and also in *Crystal Data* by Donnay *et al.*, published by the American Crystallographic Association, 1963, available from Polycrystal Book Service, G.P.O. Box 620, Brooklyn 1, N.Y. The problem will not be quite as simple for the noncubic crystals.

5 *The Thirty-Two Point Groups: Crystal Classes*

In Chapter 2 we reviewed the evidence that crystals are made up of exceedingly small structural units, repeated, side by side, indefinitely in all directions. A perfect crystal is a *homogeneous* body. Any small bit of it is just like every other small bit of it, and as we saw in Chapter 2, this places some restrictions on the kinds of symmetry it can have, with the result that there are just 32 possible crystallographic point groups.

One might have been able to arrive at this conclusion by

examining the symmetry of hundreds of crystals and finding that only 1-, 2-, 3-, 4-, and 6-fold rotation axes occur in their symmetry. One could then try to determine what it was in the nature of crystals that resulted in this limitation of their symmetry. The repetitive-unit structure of crystals might have been discovered in this way, but this was not the way it happened.

In 1830 a crystallographer named Hessel investigated the possible types of symmetry for a solid figure bounded by plane faces, a purely mathematical study. Being aware of the Abbé Haüy's Law of Rational Indices, the building-block law, he considered what types of symmetry would remain if he excluded from his figures all those that did not obey this law. He derived 32 types, and having confidence that Haüy's law represented the truth about crystals, Hessel stated that these 32 symmetry groups were the only possible ones for crystals. This conclusion was published in his book, *Krystallonomie und Krystallographie,* in Leipzig in 1831. (If astronomy, why not crystallonomy? The suffix *-onomy* is derived from the Greek verb *nemein,* meaning to distribute, arrange.) Curiously, this work appears not to have been read by other students of natural science at the time, so that Bravais in 1849 and Gadolin in 1867 repeated the derivation and came up with the same 32 point groups. Gadolin published his work both in an article in a Finnish scientific journal and in book form in Helsinki. Bravais's and Gadolin's work was well known before the crystal physicist Leonhard Sohncke rediscovered Hessel's earlier book in 1891.

The 32 crystallographic point groups are tabulated in this chapter in two different ways. In Table 5-1 the point groups are listed by symbols, and the symmetry elements of each group are tabulated. In Table 5-2 a perspective diagram of the symmetry elements of each point group is given and beside it, the stereographic projection of the symmetry elements and of a "general form." Under each of these are the symbols of the point groups. A great deal of information is condensed into these tables, and some surprising relationships are to be discovered in them.

The symbols of the point groups given in Table 5-2 are, first, those used by Schoenflies, a German mathematician-crystallographer, and second, those of C. Hermann of the University of Stuttgart and Ch. Mauguin of the University of Paris. The

TABLE 5-1 *The 32 Crystallographic Point Groups*

(Numbers in parentheses refer to axes inherent in other axes present.)

System	Point Group Symbol	How many of each kind of symmetry element?									
		m	2	3	4	6	$\bar{1}$	$\bar{2}$	$\bar{3}$	$\bar{4}$	$\bar{6}$
Triclinic	1										
	$\bar{1}$						1				
Monoclinic	2		1								
	m	1						1			
	$2/m$	1	1				1	1			
Orthorhombic	222		3								
	$mm2$	2	1					2			
	mmm	3	3				1	3			
Tetragonal	4		(1)		1						
(Every point	$\bar{4}$		(1)							1	
group has one	$4/m$	1	(1)		1		1	1		1	
4 or $\bar{4}$ axis.)	422		4+(1)		1						
	$4mm$	4	(2)		1			4			
	$\bar{4}2m$	2	2+(1)					2		1	
	$4/mmm$	5	4+(1)		1		1	5		1	
Trigonal	3			1							
(Every point	$\bar{3}$			(1)			(1)		1		
group has one	32		3	1							
3-fold axis.)	$3m$	3		1				3			
	$\bar{3}m$	3	3	(1)			(1)	3	1		
Hexagonal	6		(1)	(1)		1					
(Every point	$\bar{6}$	1		(1)				(1)			1
group has one	$6/m$	1	(1)	(1)		1	1	1	1		1
6 or $\bar{6}$ axis.)	622		6+(1)	(1)		1					
	$6mm$	6	(1)	(1)		1		6			
	$\bar{6}m2$	4	3	(1)				4			1
	$6/mmm$	7	6+(1)	(1)		1	1	7	1		1
Cubic	23		3	4							
(Every point	$m3$	3	3	(4)			(1)	3	4		
group has four	432		6+(3)	4	3						
3-fold axes.)	$\bar{4}3m$	6	(3)	4				6		3	
	$m3m$	9	6+(3)	(4)	3		(1)	9	4	3	

NOTE: $m = \bar{2}$.

TABLE 5-2

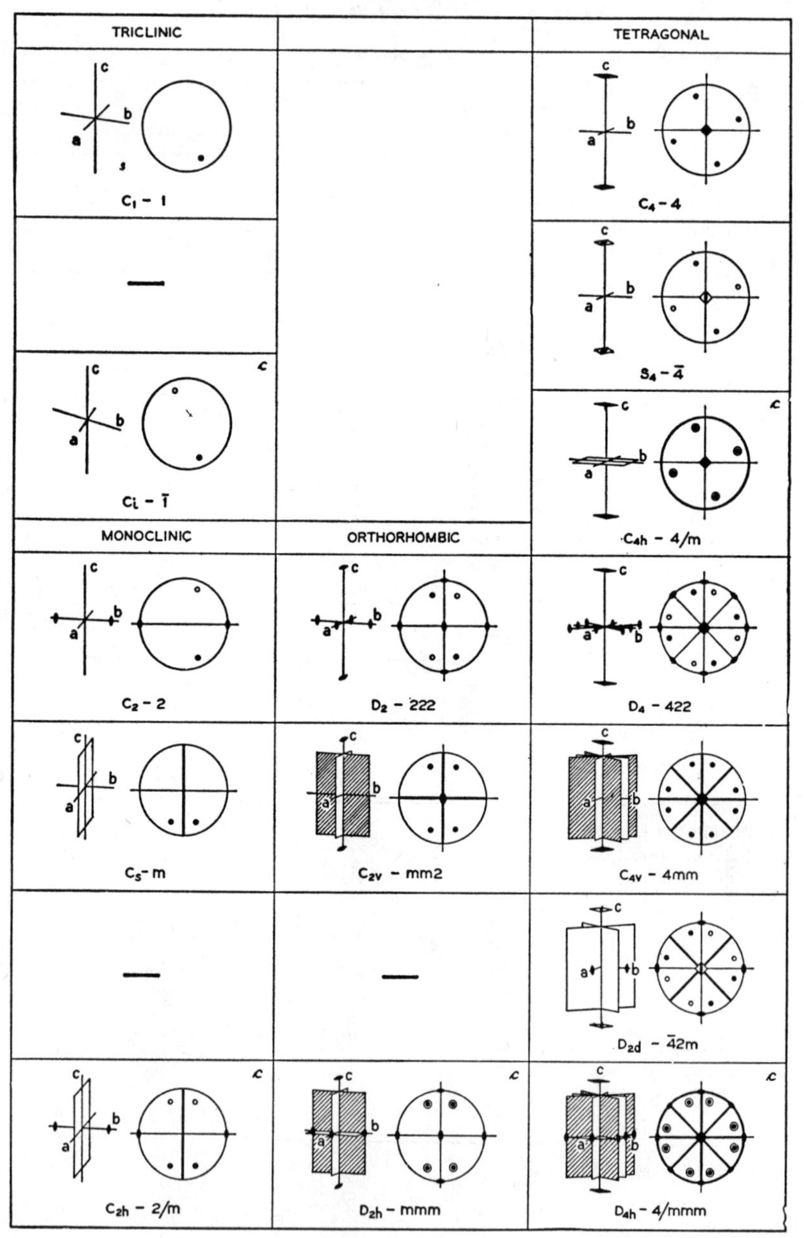
TRICLINIC
TETRAGONAL
$C_1 - 1$
$C_4 - 4$
$S_4 - \bar{4}$
$C_i - \bar{1}$
$C_{4h} - 4/m$
MONOCLINIC
ORTHORHOMBIC
$C_2 - 2$
$D_2 - 222$
$D_4 - 422$
$C_s - m$
$C_{2v} - mm2$
$C_{4v} - 4mm$
$D_{2d} - \bar{4}2m$
$C_{2h} - 2/m$
$D_{2h} - mmm$
$D_{4h} - 4/mmm$

TABLE 5-2 (cont.)

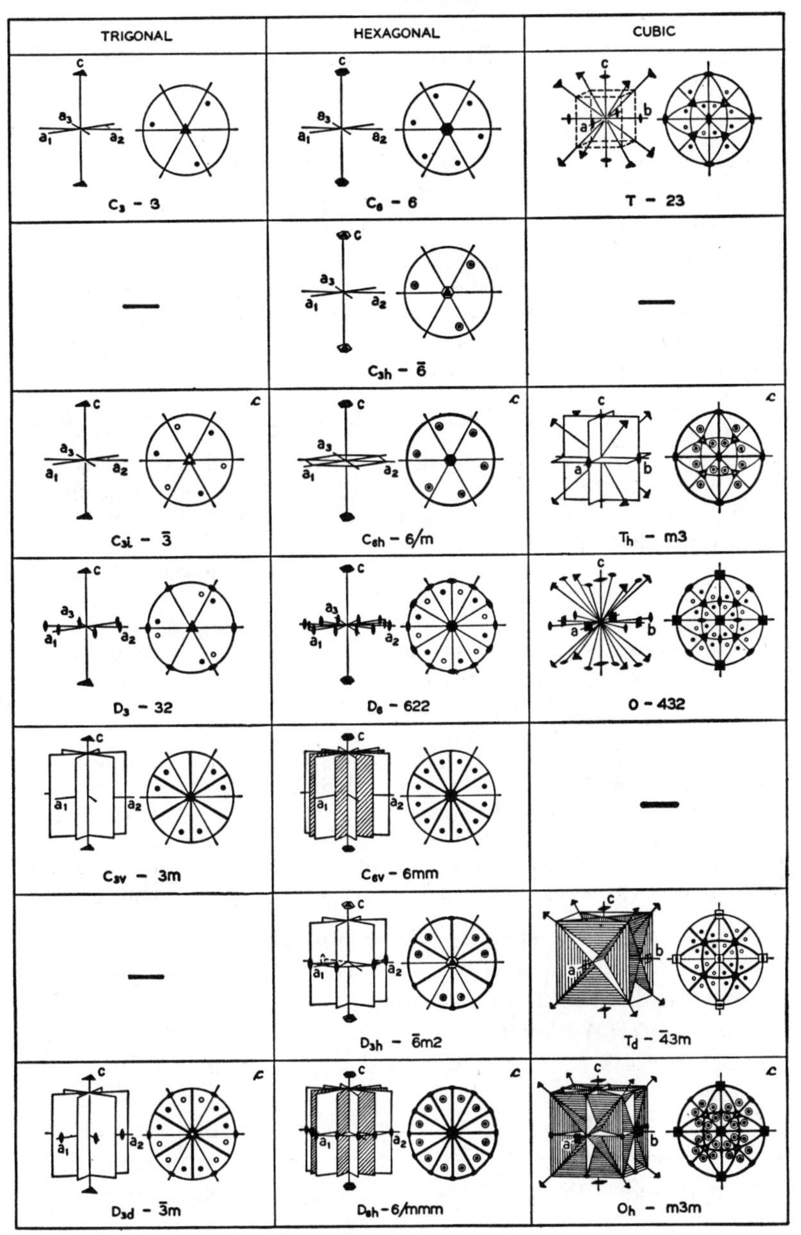

Schoenflies symbols are less commonly used, but they are still often used and are therefore included in the table for reference. The Hermann-Mauguin symbols give symmetry information as follows:

m Mirror plane
2, 3, 4, 6 Rotation axis (2-fold, 3-fold, etc.)
$\bar{1}, \bar{2}, \bar{3}, \bar{4}, \bar{6}$ Inversion axis (1-fold [= center of symmetry], 2-fold [= mirror normal to the $\bar{2}$ axis], 3-fold, etc.)

(Just as $\bar{2}$ signifies rotation of 360°/2 combined with inversion through a center-point, so $\bar{1}$ signifies rotation of 360°/1, the identity operation which just gets you back where you started, plus inversion through a center point—i.e., $\bar{1}$ signifies only inversion through a center point, indicated in Table 5-2 by *c* in upper corner of box.)

$\frac{2}{m}$ or 2/*m* ("2 over *m*") 2-fold axis with mirror plane normal to it.

*mm*2 Two mirror planes intersecting in a 2-fold axis. Sometimes written as 2*m* or *mm*. As we saw in Figs. 4-20, 4-21 and 4-22, either 2*m* or *mm* (two intersecting mirror planes) will lead you to the same final result: *mm*2.

4/*m mm* 4-fold axis with mirror plane normal to it and four mirror planes containing it. This is the highest tetragonal symmetry and is illustrated in Fig. 4-12. The four two-fold axes normal to the 4-fold axis need not be mentioned, since you cannot avoid acquiring them. This symbol is usually "closed up," written 4/*mmm*, but of course it is obvious that only one of the five mirrors can be normal to the 4-fold axis.

432 This refers to a point group that has 3-fold axes which are not parallel to *c*. This can only occur in the cubic system.

The first number of a symbol refers to an axis parallel to *c*, except in the monoclinic system where the unique axis is *b* and in the orthorhombic system where the order is *a, b, c*. The second number refers to axes normal to the first, except in the cubic system.

The cubic system poses a special problem because, unlike the other systems, it has symmetry axes which are neither along nor normal to the crystallographic axes. These are the 3-fold axes along the ⟨111⟩ directions, present in every cubic point group. Therefore a 3 is always the second item in a cubic point-group symbol.

Fig. 2-10, which gave the dimensional relationships of the seven types of unit cells, and Fig. 2-24, the fourteen space lattices, both included the word *rhombohedral,* which is missing from

Tables 5-1 and 5-2. A word of explanation is in order. All crystals in which the unique symmetry axis is a 3-fold axis are trigonal crystals. [Note that this excludes cubic crystals, where there are four 3-fold axes, not a unique one. It also excludes $\bar{6}$, a point group which could be called $3/m$ (Fig. 4-23).] The smallest unit cell of some trigonal crystals is rhombohedral in shape, whereas for others, even in the same point group, the smallest unit cell is "hexagonal," i.e., the shape of Fig. 2-10d. "Rhombohedral" was formerly used for all trigonal crystals, but is now reserved for those with a rhombohedral unit cell (Fig. 2-10e). The term "rhombohedral" used in earlier texts thus did not have the same significance that it has today.

For point-group symmetry the term *trigonal* is unambiguous as defined above.

Symbols used in Table 5-2:

Rotation axis 2, 3, 4, 6, respectively.
Inversion axis $\bar{3}$, $\bar{4}$, $\bar{6}$, respectively.

Crystallographic axes are marked at the rim of the projection by short radial straight lines. The symmetry planes and the projection of the general form have been discussed in Chapter 4.

Axes that are inherent in other axes are indicated in Table 5-1 in parentheses. For example, as pointed out in Chapter 1 with reference to the card table, a 2-fold axis is always inherent in a 4-fold axis. If you have a 4-fold axis, it causes repetition every 90°, and $2 \times 90° = 180°$, which is the repetition period of a 2-fold axis.

There are some amusing relationships among these axes. For example, 3 is inherent in $\bar{3}$, as you can see by Table 5-2, where the projection of the general form for point group $\bar{3}$ also has the symmetry 3. However, 4 is not inherent in $\bar{4}$. What about $\bar{6}$ and $\bar{2}$? You will not find $\bar{2}$ in Table 5-2, but you will find that it is equivalent to m, which has arbitrarily been chosen as the symbol for the point group instead of $\bar{2}$. Either would have been satisfactory.

The term *crystal class* refers to a class of crystals which have a given point-group symmetry. There are thus 32 crystal classes. In the past the crystal classes were given names which were often long and cumbersome, and different authors assigned different names to the same class. For example, the cubic class with the point-group symmetry $m3$, the symmetry of the mineral pyrite or fool's gold (FeS_2), has been variously called by the following names: Diakisdodecahedral, Didodecahedral, Pentagonal Hemihedral, Diploidal, and Tesseral Central. Since some authors listed the classes beginning with the cubic classes, while others began

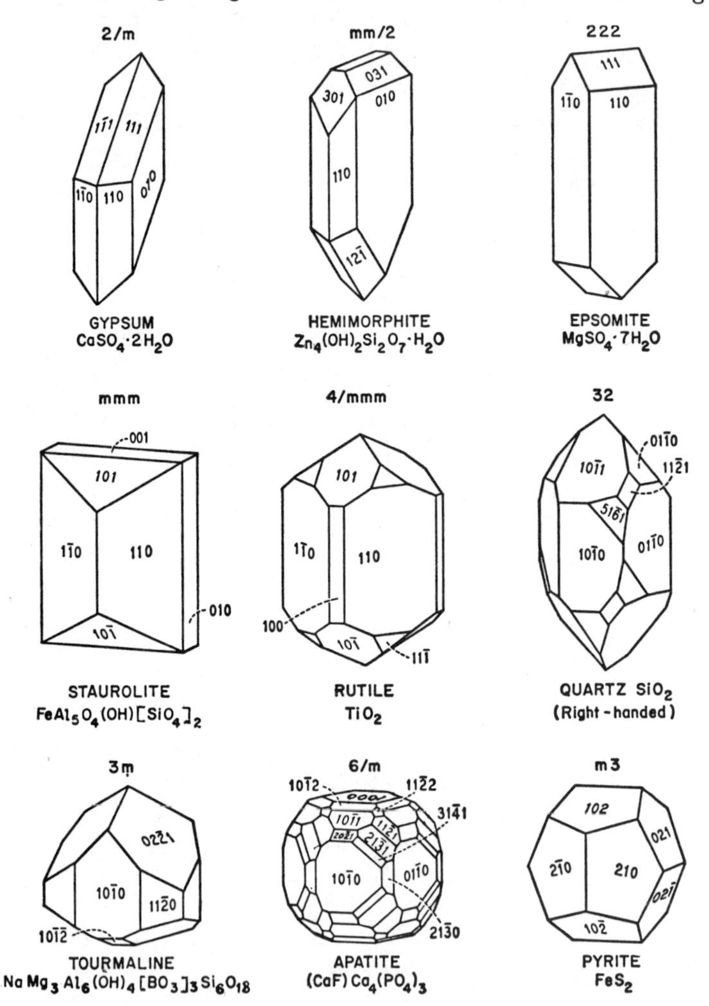

FIG. 5-1

with the triclinic, the numerical order could not be used for identification. Recently the symbol for the point-group symmetry has generally been used to designate the crystal class.

In spite of this shifting nomenclature, there is no doubt about the classes themselves. Since Hessel derived them in 1830, no writer has disagreed with the choice of the 32 symmetry groups. You can recognize them by their symmetry in the oldest texts on crystals, even though their names may be jaw-breakingly unfamiliar.

Fig. 5-1 shows crystals of a number of different substances, illustrating various point groups. Compare Table 5-2.

* * *

Problem: Assign the proper point-group symbol to:

1. A simple four-legged table like that in Fig. 1-4.
2. A brick (Fig. 1-7).
3. A disphenoid (Fig. 1-11).
4. A common hammer.
5. The scheelite crystal in Fig. 4-16.
6. The garnet crystal in Fig. 4-13.

To what crystal system would each belong if it were a crystal? (Answers are given at the end of Chapter 6.)

* * *

Answers to questions at the end of Chapter 4

1. 2. 3. 4.

5. No. 3 and No. 4: each has a center of symmetry.

6 *The Crystalline State*

Up to this point, we have been examining the orderly nature of crystals, their repetitious structure and their symmetry. Beyond this point we are going to see some of the results of this orderliness in the physical properties of crystals. At this midpoint is a basket chapter which holds an assortment of facts about crystals that will be useful to know.

Nearly all solid substances have an orderly arrangement of their constituent atoms, some more orderly than others. The structure of textile fibers has been studied from their x-ray diffraction patterns, and even a match stick gives a good pattern. The textile fibers have much better order along the fibers, as you might expect, than across them. Whether you call these "crystals" or not depends upon how exclusive you want the definition to be. The fiber is "a solid composed of atoms arranged in an orderly, repetitive array." It doesn't have very good three-dimensional order and it doesn't develop shiny faces like those of the topaz crystal, but crystallographers study it and report their results at meetings of the American Crystallographic Association.

Teeth are made up of small crystals of calcium phosphate, and eggshells are made of calcium carbonate crystals. These are good proper crystals with three-dimensional order, but they are very small and all intergrown, one against the other; their crystal faces have not developed because neighboring crystal grains got in the way. Such a polycrystalline (many-crystal) mass might have grain boundaries such as those shown in Fig. 6-1, which is a thin section of Carrara marble magnified 18 times. Here the crystals are calcite crystals ($CaCO_3$), and the straight lines in many of the grains mark cleavage planes which will be discussed in the next chapter. These planes are parallel to $\{10\bar{1}1\}$ in this crystal, whose point group is $\bar{3}m$. Within any single crystal we can see that the orderly array of the structure is maintained, since these planes are straight and continuous. The neighboring crystal may have a very different orientation, but inside its boundary it is homogeneous in its orderly array. This array is exactly the same as that in the crystal shown in Plate II. The fact that the very small crystal in the Carrara marble has grown around and against its neighbors, and therefore shows no faces, has in no way affected the orderly accumulation of the carbon, oxygen, and calcium atoms that make up the structure of calcite.

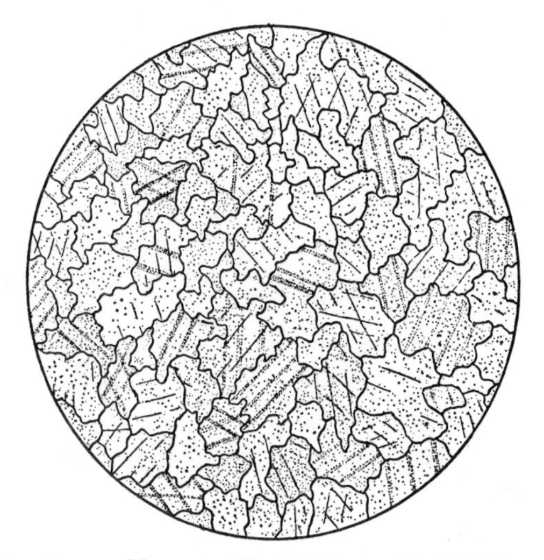

FIG. 6-1 Carrara marble, magnified 18 times. (From *Metamorphism*, by Alfred Harker. By permission of Methuen & Co. Ltd.)

Many of the polished rocks used for decoration on buildings are quite coarsely crystalline, so that you can see the grain boundaries between the crystals without a microscope.

Plate III (2) is a photomicrograph of a piece of polished brass which has been etched with acid. The dark and light areas are due to the acid acting at a different rate on the differently oriented crystals. Here again, the irregular boundaries are grain boundaries between crystals. The straight boundaries are boundaries of a different kind.

These straight boundaries are due to *twinning*, the name given to a "mistake" the crystal makes in growing. In twinning, the later-grown part is not parallel to the earlier-grown part

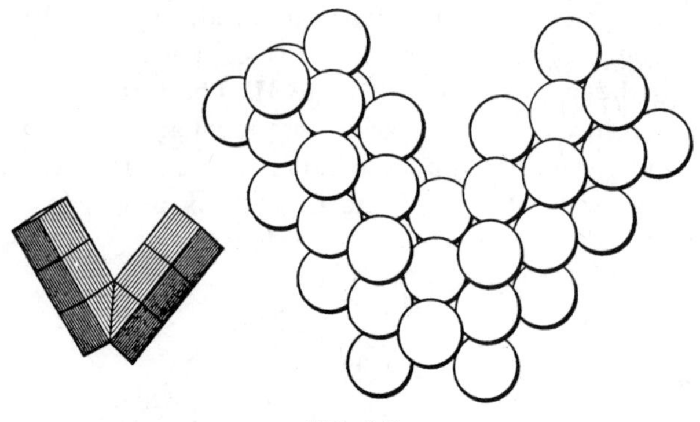

FIG. 6-2

(derivable from it by pure translation), but is related to the earlier-grown part by some symmetry operation which is not a symmetry operation of the single crystal. A few examples will make this geometrical relationship clear. Fig. 6-2 is a sketch of a

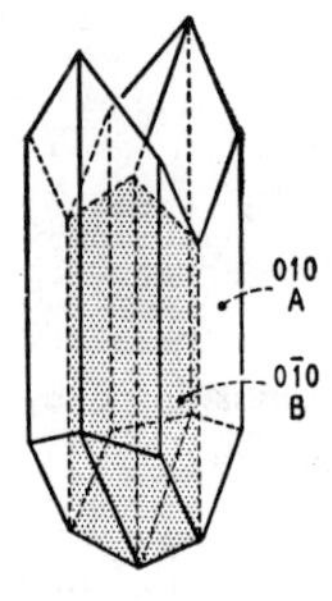

GYPSUM

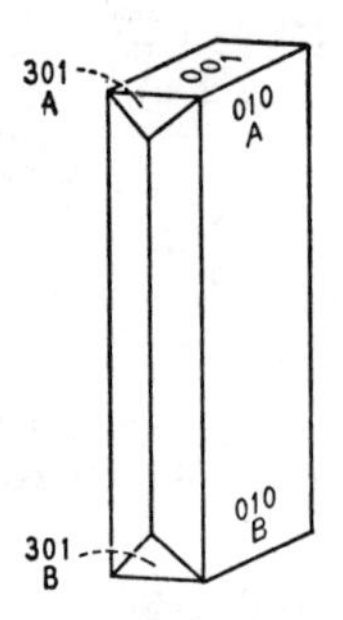

HEMIMORPHITE

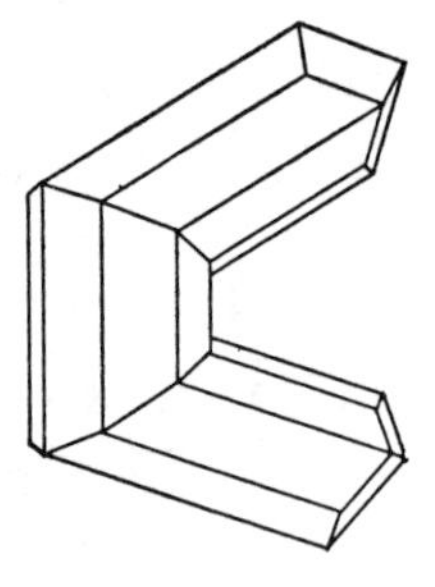

RUTILE

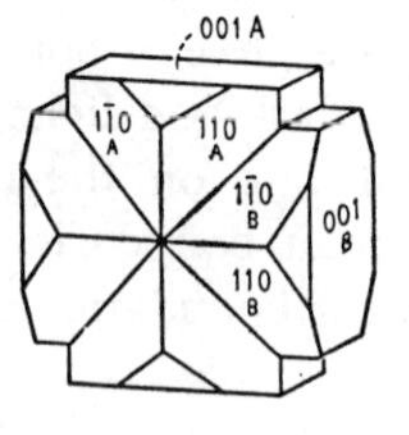

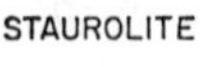

STAUROLITE

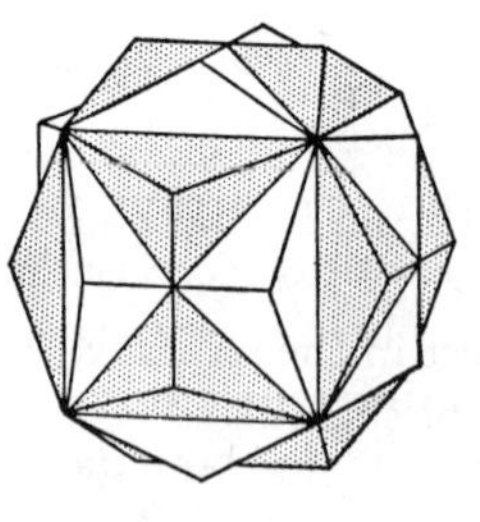

PYRITE

FIG. 6-3

structure model of a twin in a face-centered cubic element such as copper. At some particular (111) plane, the arriving atoms took positions which were the mirror image of those on the other side of the (111) plane. Since this plane is not a mirror plane for this crystal, these were the wrong positions. However, the crystal continued to grow in the new orientation, the new part being related to the old part by a mirror plane, known as the twinning plane.

In Fig. 6-3 some examples of twinned crystals are shown. Compare them with Fig. 5-1. One individual of the twin is always related to the other by a symmetry operation, and the twins share a plane of atoms that fits right into the structure of each. It is this that distinguishes the twin from a couple of crystals that just happened to grow together. Such crystals are shown in Plate II. They are not twins, but just intergrown crystals of calcite. Like neighboring grains in the Carrara marble of Fig. 6-1, they bear no special orientational relationship to each other.

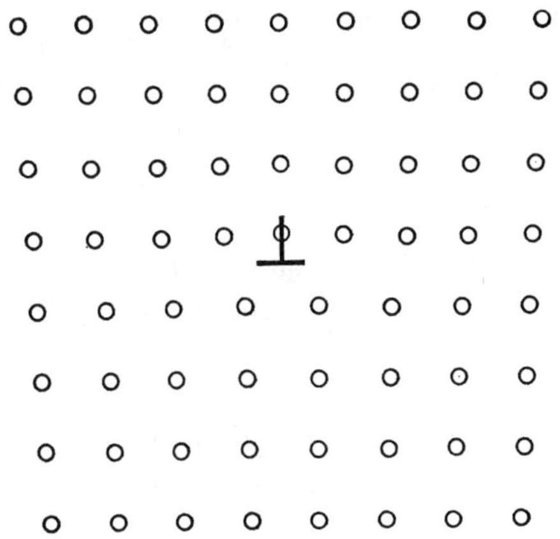

FIG. 6-4

FIG. 6-5

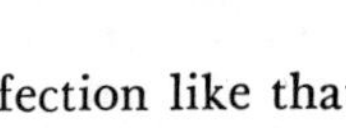

In the process of growth of a crystal, an imperfection like that shown in Fig. 6-4 may occur. The inverted T calls attention to the defect in the structure. Such a defect is called a dislocation. If a series of such defects occurs one above the other, as in Fig. 6-5, the array of atoms on one side of the defects will not be parallel to the array on the other side. The angle between them will depend upon how closely spaced the dislocations are. The boundary between them is a low-angle (small-angle) grain boundary. Many such low-angle grain boundaries may occur in the course of crystal growth, as shown in Fig. 6-6. This figure is from

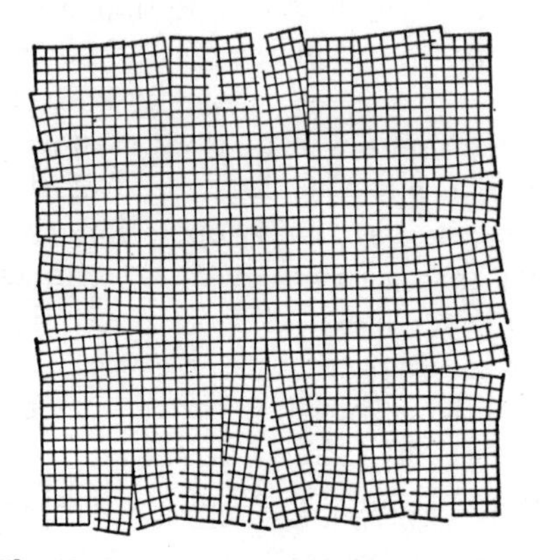

FIG. 6-6 (From *The Lineage Structure of Crystals*, Zeitschrift für Kristallographie, vol. 89, 1934, p. 195, by permission of the author, M. J. Buerger.)

a paper published in 1934 by M. J. Buerger, who called this particular kind of imperfect crystal growth "lineage." Is a crystal that grows this way a "single crystal"? Some would say yes: it is certainly very different from the polycrystalline Carrara marble of Fig. 6-1. Some would say no: there are boundaries in it that separate nonparallel regions, so that not every unit cell in it can be derived from any other by straight translation. It is a crystal with lineage; a crystal with low-angle grain boundaries.

* * *

Answers to questions at the end of Chapter 5

1. Table: $4mm$ (tetragonal).
2. Brick: mmm (orthorhombic).
3. Disphenoid: $\bar{4}2m$ (tetragonal).
4. Hammer: m (monoclinic).
5. Scheelite crystal: $4/m$ (tetragonal).
6. Garnet crystal: $m3m$ (cubic).

The symmetry elements of the disphenoid (No. 3 above) were discussed in Chapter 1 to illustrate the operation of the 4-fold inversion axis, $\bar{4}$. Attention was not called to the fact that it also has three 2-fold axes through the centers of its opposite edges. One of these coincides with the $\bar{4}$ axis and is inherent in it. If you placed a thumb and finger at these points and rotated the sphenoid of Fig. 1-11 through 180° around such an axis, its new position would be indistinguishable from its original position. The existence of the two 2-fold axes normal to the 4-fold inversion axis would also be discovered by plotting the stereographic projections of the faces of the sphenoid relative to the 4-fold inversion axis and mirror planes, as in Fig. 6-7.

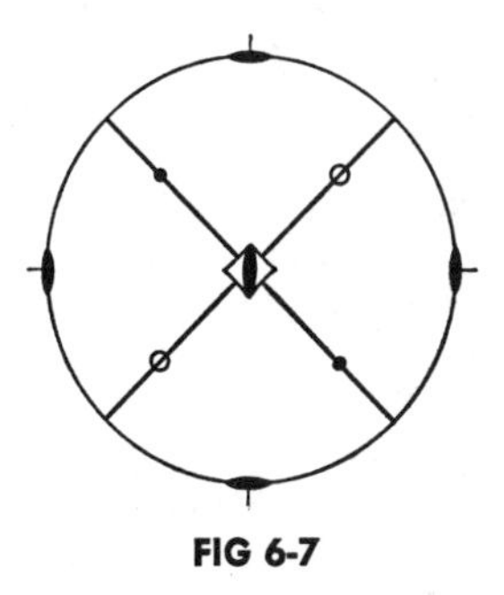

FIG 6-7

With respect to the perfection of their symmetry, the six objects above fall into three grades. We might distinguish them arbitrarily as follows: *A*, perfectly exemplary of the point group listed; *B*, almost perfectly exemplary; and *C*, only roughly exemplary in external shape. Can you determine which objects belong in each category?

Since the disphenoid is a geometrical object constructed from our imagination, we can imagine it to be as perfect as we please, so it belongs in grade *A*. The two crystals, scheelite and garnet, almost certainly have minor imperfections in the arrangement of their atoms, but except for these occasional flaws, their structure throughout has the symmetry indicated by their external appearance, i.e., by the point group given. So the crystals belong in grade *B*. The table and hammer handle are made of wood fibers and the brick and the head of the hammer are polycrystalline, looking like Fig. 6-1 under the microscope; their internal structure certainly bears no symmetry relation to their external shape, which was formed by manufacturing processes. They belong to grade *C*.

* * *

Sources of crystals with well-developed faces

In many parts of the country there are well-organized mineralogical and geological clubs that go on mineral-collecting trips. Often they are associated with the local science museum. Contact with such a group may be a good source of crystals.

There are many companies that sell specimens of crystals with well-developed faces. Two of the largest that will send illustrated catalogues on request are Ward's Natural Science Establishment, Rochester, New York, and Eckert Mineral Research, 110 East Main St., Florence, Colorado.

Sodium chloride crystals may be found in the nearest salt shaker. By dissolving these in water and letting the solution evaporate, you can grow bigger ones. For further information on growing your own crystals, see *Crystals and Crystal Growing* by A. N. Holden and P. Singer, Doubleday-Anchor, Garden City, N.Y., 1960.

Synthetic crystals of a number of substances are available from several firms whose advertisements may be found in the technical journals. They are generally purer and more expensive than most natural crystals.

7 *The Relation Between the Symmetry of a Crystal and the Symmetry of Its Physical Properties*

In a footnote in Chapter 2 an accident was mentioned as the cause of Abbé Haüy's becoming a crystallographer. An account of the accident was promised for Chapter 7. When René Just Haüy was teaching at Lemoine College in France, he began to devote all his spare time to the study of botany, but a good friend of his was a mineral collector and, probably because of trips he had taken with his friend, Haüy had a small collection himself. One day his friend showed him a particularly fine specimen of calcite. From the description that has come down to us, it must have looked rather like the specimen in Plate II of this book.

While Haüy was examining the treasured specimen, he dropped it. One of the large crystals broke off, and his good friend let him have it for his collection! But Haüy had noticed a surprising thing. The broken surface was a smooth bright face, and he was curious to see whether the same sort of face would result if he broke it again.

"The prism had a single fracture along one of the edges of the base," he wrote later, "by which it had been attached to the rest of the group. Instead of placing it in the collection which I was then making, I tried to divide it in other directions. . . ." He found that there were three directions in the crystal along which he could break it and get a bright plane surface like the first one. Such an easy plane of parting in a crystal is called a *cleavage plane*. Cleavage is the tendency of the crystal to come apart between planes of atoms when struck or pulled apart. Mica has excellent cleavage and can be pulled apart in thin sheets. Some crystals do not exhibit cleavage at all. In order to have an easy cleavage, the bonding between the atoms in the direction normal to the cleavage plane must be weaker than in other directions.

In calcite the cleavage plane is $\{10\bar{1}1\}$, expressed on hexagonal axes. The point group of calcite is $\bar{3}m$, and therefore the complete form $\{10\bar{1}1\}$ has six faces, in three parallel pairs, as shown in stereographic projection in Fig. 7-1. [Faces such as $(0\bar{1}11)$ and $(01\bar{1}\bar{1})$ are of course parallel.] Plate VI(2) is a photograph of a cleavage rhombohedron of calcite with $\{10\bar{1}1\}$ faces. If this piece of calcite were broken again and again into very small fragments, every fragment would be bounded by $\{10\bar{1}1\}$ faces, and therefore the angular relations between the faces would be the same in every fragment (see Plate VI(3)).

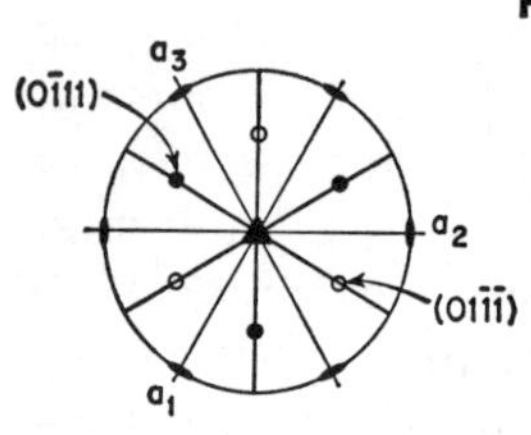

FIG. 7-1

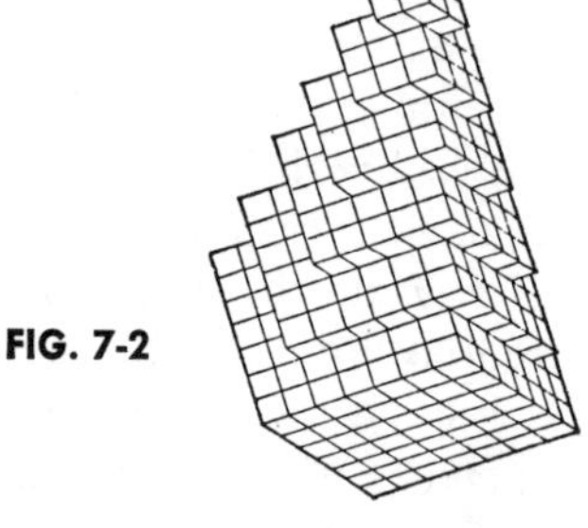

FIG. 7-2

When Haüy observed that the angular relations between the cleavage faces were the same in every fragment (as in Plate VI(3)), he erroneously concluded that he could break the crystal into the small pieces which were the ultimate building blocks of which it was constructed: ". . . and I succeeded after several trials," he wrote, "in extracting its rhomboid nucleus." In spite of this error, he did have the concept of a repeat unit, and he showed that he could reconstruct the original sharply pointed crystals, with the proper interfacial angles, by stacking up the little cleavage blocks which he called "molécules intégrantes." One of his figures is reproduced here as Fig. 7-2.

The excitement of this discovery resulted in his forsaking the study of botany for crystallography.

Cleavage, like all other properties of a crystal, is the result of the nature and arrangement of the atoms in the crystal. As soon as you know the symmetry of the crystal structure, you know the symmetry its properties must have.

Given a crystal with point-group symmetry $\bar{3}m$ and cleavage parallel to $(10\bar{1}1)$, you know that the $\bar{3}$ axis means that the other planes of Fig. 7-1 will also be cleavage planes. Since three differently oriented planes will bound a solid if they do not all belong to the same zone, all surfaces of a broken piece of this crystal will be cleavage planes.

Plate VI(3) shows a number of such broken pieces of calcite. Some are long and narrow, others more nearly equidimensional, depending on where the break happened to occur. Since the cleavage depends on the arrangement of the atoms, there are potential cleavage planes throughout the crystal as closely spaced as the repeat unit of the structure.

Plate VI(1) is a sketch of a model of the structure of calcite. The surfaces of the model show the orientation of the cleavage planes in the calcite structure. Compare the shape of the model with that of the piece of calcite in Plate VI(2). Since the unit cell is a few Ångström units on an edge (a few $\times$ 10^{-8} cm), there are about 10^8 potential cleavage planes in one direction in a piece of calcite a centimeter long.

What is the symmetry of the broken bits of calcite in Plate VI(3)? If we think only of their external shapes, many of them would appear to have lost the $\bar{3}$ axis, but each bit of calcite is still composed of calcium, carbon, and oxygen atoms arranged in an orderly structure with point-group symmetry $\bar{3}m$. One might say that, since each unit cell has this symmetry, there are millions of $\bar{3}$ axes and mirror planes, all self-parallel, in any bit of the crystal. The orientation of the cleavage planes will always be in accordance with this symmetry, but, just as in the case of the growth forms in Figs. 4-2 and 4-3, the exterior dimensions of any cleavage bit will depend on the conditions of its formation.

The common mica, muscovite, is a monoclinic crystal with point-group symmetry $2/m$ and cleavage parallel to $\{001\}$. How many differently oriented cleavage planes has it? From Table 5-2 we see that the mirror plane in the monoclinic system is normal to the *b* axis and the 2-fold axis is taken as the *b* axis, as shown in Fig. 7-3. Since the action of a mirror on a plane normal to it is

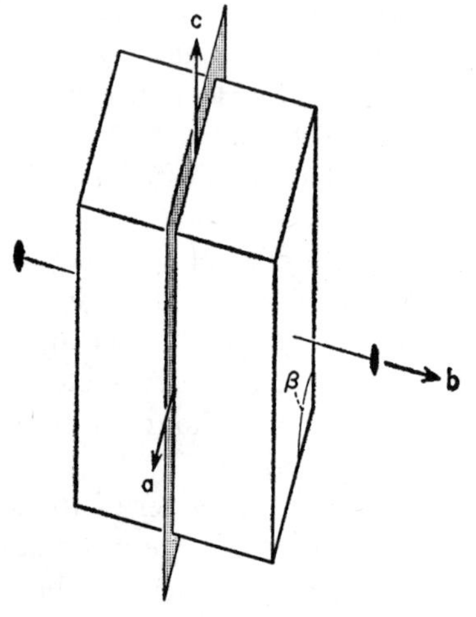

FIG. 7-3

just to reflect it into itself, the action of the mirror on the plane (001) in Fig. 7-3 will not generate any additional planes. However, the 2-fold axis parallel to (001) indicates that $(00\bar{1})$ is a symmetrically equivalent plane, that is, the form {001} in muscovite mica consists of (001) and $(00\bar{1})$. Since these two faces are parallel, the cleavage in mica has only one orientation. Unlike calcite, therefore, it cleaves in sheets, not in chunks.

Now suppose we consider some other physical properties. Take, for example, the property of elasticity, thc ability of a substance to change its dimensions elastically in response to an applied stress. (An "elastic" change refers to a change which exists only as long as the stress is applied. When the stress is removed, the object reverts to its original dimensions.) Each atom in a crystal is where it is as a result of the balance of attractive and repulsive forces which it and its neighbors are exerting on each other. If we apply a stress to a crystal, we change the balance of forces, and the positions of the atoms change in response—i.e., there is a resulting strain (dimensional change).

If we apply a compressive stress along any one of the ⟨100⟩ directions in sodium chloride (Fig. 4-27), we will be tending to shorten mostly the distance between unlike atoms, whereas a compressive stress along a ⟨101⟩ direction (the face diagonal in Fig. 4-28) will tend to shorten mostly the distance between like atoms. It is therefore not surprising that the elastic constants (the quantities expressing the amount of strain caused by a given stress: the "yield," or *compliance*) are different in these two directions.

If we drew an arrow in the [100] direction with its length in millimeters equal to the magnitude of the elastic constant for compression in that direction and a similar arrow of appropriate (different) length in the [110] direction, these arrows would represent the elastic constants in these two directions. To show the complete picture of the way the elastic constants vary for all directions in the crystal, we would have to have an infinite number of arrows, radiating from the center. Their tips would define a surface, an envelope, touching every arrow tip, which would represent the variation of elastic constant with direction. We could, if we chose, draw arrows representing stiffness to stress, the inverse of compliance. Fig. 7-4 shows the resulting surface for Young's modulus (which is a measure of elastic *stiffness* to compression in various directions) in quartz. By examining the symmetry of this surface and the symmetry of the quartz crystal in Fig. 5-1, could you say approximately which direction in the crystal is the direction with the largest Young's modulus, the stiffest direction, the direction that would deform least when a compressive stress was applied along it?

(As always, in speaking of properties of crystals, when we speak of the value of a property "in a direction" we mentally include those other directions that are crystallographically equivalent to this one, since the property will have the same value in those directions because of symmetry.) Since the angular position of the longest arrow (outermost point) of the elasticity surface in Fig. 7-4 is not given, you can, of course, say only approximately what direction in quartz it corresponds to. The symmetry elements of the elasticity surface must coincide in space with the symmetry elements of the crystal faces, since both result from the symmetry of arrangement of the crystal's atoms.

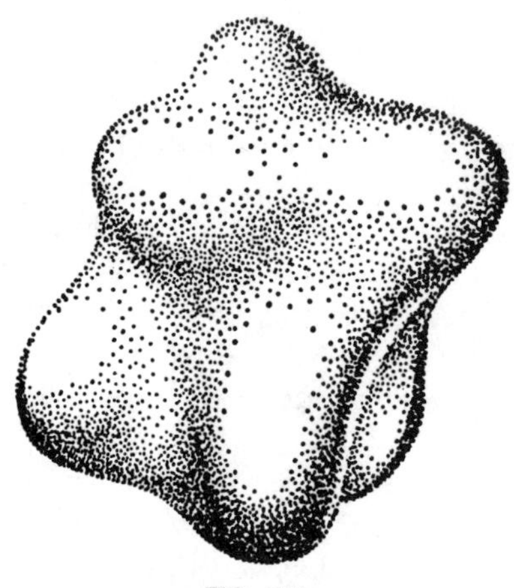

FIG. 7-4

Note that the elasticity surface in Fig. 7-4 has a center of symmetry, although the quartz crystal does not. The surfaces representing many properties have symmetries higher than that of the crystal. Such is the case, for example, in linear thermal expansion, the change in the dimension of an object along some particular direction as a result of a change in its temperature. If a cubic crystal is heated, it expands uniformly in all directions: the expansion surface is a sphere. If you cut a perfect sphere from a cubic crystal, it would remain perfectly spherical as you heated or cooled it. In crystal systems where one axis is unique (unlike any of the others) but the two normal to it are symmetrically equivalent to each other, the surface for linear thermal expansion is an ellipsoid of revolution. An ellipsoid of revolution is a solid figure formed by rotating an ellipse around one of its two axes. If it is rotated around its major axis, the resulting solid will be longest along its rotation axis, like Fig. 7-5. If it is rotated around its minor axis, it will be shortest along its rotation axis, like Fig. 7-6. The lines on both figures represent various positions of the rotating ellipse. All sections normal to the axis of revolution are, of course, circular.

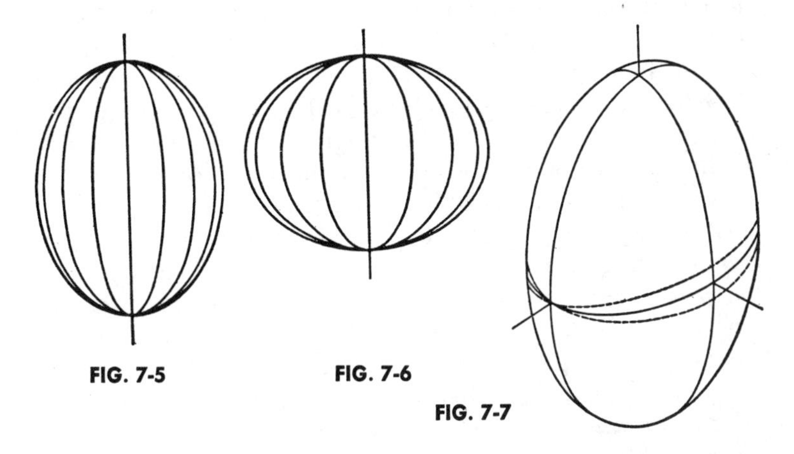

FIG. 7-5 FIG. 7-6 FIG. 7-7

In crystals where no two axes are crystallographically equivalent, the vector surface for thermal expansion is a triaxial ellipsoid. In the triaxial ellipsoid (Fig. 7-7) there are three mutually perpendicular axes which are, in general, unequal in length. There are only two circular sections (dotted in Fig. 7-7) that contain the axis of intermediate length. The angles at which they intersect with the long and short axes depend upon their relative lengths. All other central sections are elliptical.*

Suppose we examine the symmetry of these solid figures, just as we did the symmetry of various solid figures back in Chapter 1. The ellipsoids of revolution, Figs. 7-5 and 7-6, have the symmetry ∞/m, $2/m$—i.e., such a figure looks the same an infinite number of times during a 360° rotation around its symmetry axis and

* The original for Fig. 7-7 was drawn by a computing machine. Miss Ruth A. Weiss of the Bell Telephone Laboratories programmed the information for the three different ellipses, mutually perpendicular, seen in perspective, and the machine calculated the curves and produced the diagram on transparent tape, from which an enlarged print was made. This print was traced by the draftsman with improved quality of line.

there is a symmetry plane normal to this axis. There is also an infinite number of symmetry planes containing the rotation axis and an infinite number of 2-fold axes normal to it. As we look over the symmetry elements of the 32 point groups in Table 5-1 or 5-2, we see that the point group of an ellipsoid of revolution includes the symmetry elements of any point group except those of the cubic system. The ellipsoids in Figs. 7-5 and 7-6 do not have the four 3-fold axes that characterize every cubic structure. Therefore no property of a cubic crystal could be represented by a surface of that shape. The point group of a sphere, with its infinite number of axes of infinite symmetry of course includes the four 3-fold axes required by cubic crystals.

When the surface representing a particular property of a crystal is spherical, as in the case of thermal expansion in cubic crystals, the crystal is said to be *isotropic* with respect to that property. The word is made up of the prefix *iso-,* from the Greek *isos,* meaning equal, and the suffix *-tropic,* from the Greek *-tropos,* meaning *turning,* originally, but now more broadly used in the sense of *turning toward,* or *responding to.* (The heliotrope is a flower which turns toward the sun, the Greek god Helios.) The thermal expansion of cubic crystals is isotropic: the crystals *respond to* heat by expanding equally in all directions. In some other properties, cubic crystals are not isotropic, that is, they are *anisotropic.* If the rate of growth of cubic crystals were isotropic, they would all be spheres!

Although the point-group symmetry of the ellipsoid of revolution includes the symmetry elements of all the point groups except those of the cubic system, not all noncubic crystals have thermal-expansion surfaces of this shape. The reason is that the triaxial ellipsoid will satisfy the symmetry of the orthorhombic, monoclinic, and triclinic crystals just as well as would the more symmetrical ellipsoid of revolution.

The symmetry of the triaxial ellipsoid is *mmm* (see Table 5-2), with three mutually perpendicular planes meeting in three 2-fold axes along its three nonequivalent axes, shown perpendicular to each other in Fig. 7-7. Clearly such a figure would not fit the symmetry of the tetragonal, trigonal, and hexagonal point groups because it has no 4-fold, 3-fold, or 6-fold axes, but it is quite compatible with the orthorhombic, monoclinic, and triclinic point groups.

As in the case of the Young's modulus figure for quartz, so here too we can see what orientation the thermal-expansion surfaces must have with respect to various crystals.

Consider first the ellipsoid of revolution, the surface representing linear thermal expansion for the tetragonal, trigonal, and hexagonal systems. The unique crystallographic axis (*c*) in each case (4-fold, 3-fold, and 6-fold, respectively) must coincide with the infinite rotation axis of the ellipsoid. No other direction in the ellipsoid has rotational symmetry any higher than 2. Whether the thermal-expansion ellipsoid is longest along its axis as in Fig. 7-5 or shortest along its axis as in Fig. 7-6 will depend on the nature of the particular substance in question, as will also the relative dimensions, the ratio of the axial diameter to the equatorial diameter.

The orientation of the triaxial ellipsoid of thermal expansion in the orthorhombic system is fixed by the fact that its three 2-fold symmetry axes, indicated in Fig. 7-7, must coincide with the three orthorhombic crystallographic axes in order that the symmetry elements of the crystal will coincide with those of this surface. This orientation does not place any restriction on which of the three ellipsoid axes (all different in length) will lie along which of the orthorhombic crystallographic axes (*a*, *b*, and *c*), so there are six possible orientations in this case. Again the relative dimensions of the thermal-expansion surface will depend on the nature of the particular substance in question.

The only restriction on the orientation of the triaxial ellipsoid in the monoclinic system is that one of its 2-fold axes will coincide with the 2-fold axis of the crystal in those classes where the crystal has a 2-fold axis, or one of its planes will coincide with the symmetry plane in the crystal where it has a symmetry plane.

In the triclinic system the triaxial ellipsoid surface may have any orientation relative to the axes of the crystal.

This discussion of the linear thermal-expansion surfaces, their symmetry, and its relation to the symmetry of the crystals whose expansion they represent has been followed in some detail because it illustrates a way in which the knowledge of the point group of a crystal tells us something about one of its physical properties which can be measured in various directions in the crystal.

Exactly the same considerations are applicable to the optical properties of crystals which are discussed in the remaining chapters.

We can see that in order to determine the linear thermal expansion of a cubic crystal along any direction at all in the crystal, we need only measure it along one direction. We can cut a long bar with its length along any direction whatever and measure the change in length per centimeter per degree of change in temperature, and because the expansion surface is spherical, we will know the change in length for any piece cut out in any other direction.

Such properties as conduction of electricity and conduction of heat also have different values according to the crystallographic direction in which the conduction is taking place. Another case of such a property will be described in Chapter 17.

For measurements of all such properties to be meaningful in terms of understanding their relation to the structure of the crystal, we must be able to *orient* the crystal—that is, to know how its crystallographic axes lie in relation to the direction in which the measurement is being made.

In some cases x-ray diffraction must be used to determine the orientation of a crystal. One of the references at the end of this book describes methods for using x-ray diffraction for crystal orientation. However, in many cases the necessary information can be determined much more quickly and easily from a knowledge of the optical properties of the crystal, the subject of the remaining chapters of this book.

* * *

Problem: How many differently oriented bars must we measure, and what must their orientation be to determine linear thermal expansion:

1. In a tetragonal crystal?
2. In an orthorhombic crystal?
3. In a monoclinic crystal?

(Answers are given at the end of Chapter 8.)

8 *The Velocity of Light in Cubic and Uniaxial Crystals: Observation Between Crossed Polarizers*

As in the case of thermal expansion, cited in the previous chapter, the surface representing the velocity of light in any cubic crystal is a sphere. For any noncubic crystal it is not.

In order to discuss light in crystals we must first consider light itself. The concept of light that one uses in a particular situation depends somewhat on the situation. In what follows it will be convenient to think of light as a wave like that in Fig. 2-16. There the plane in which the waves are oscillating (from crest to trough) is the plane of the paper, and the oscillation (or vibration) direction is up and down, normal to the propagation direction which is toward the right. (This "vibration direction" is direction of the "electric vector" in the terminology of electromagnetic radiation.)

In a beam of unpolarized light, each ray is vibrating normal to the propagation direction, and any orientation of the vibration plane around the beam axis is possible. If, from such a beam, those rays are selected whose vibration directions are all parallel to each other, the resulting light is said to be *polarized.* If the light has all wavelengths of the visible spectrum (ranging in length roughly from $\lambda = 4000$ to $\lambda = 7000$ Å.), then it is white light. If the wavelengths of the rays are limited to a narrow range (ideally, a single value), the light is said to be *monochromatic* (from the Greek *monos,* one; *chroma,* color).

The easiest source of light confined to a narrow band of wavelengths is the flame from some substance that has been wet with a solution of sodium chloride (table salt) and then dried. The heated sodium atoms impart an orange-yellow color to the flame. This light, especially strong for a narrow band of wavelengths near 5,890 Å., is commonly called *sodium light.*

Now let us examine the surface representing the velocity of light in crystals.

Consider the simpler cubic case first. If we imagine a source of light at a center point within the crystal, turned on at some instant in time, we find that after an extremely short interval of time the light will have progressed the same distance in every direction through the crystal (Fig. 8-1), just as it would

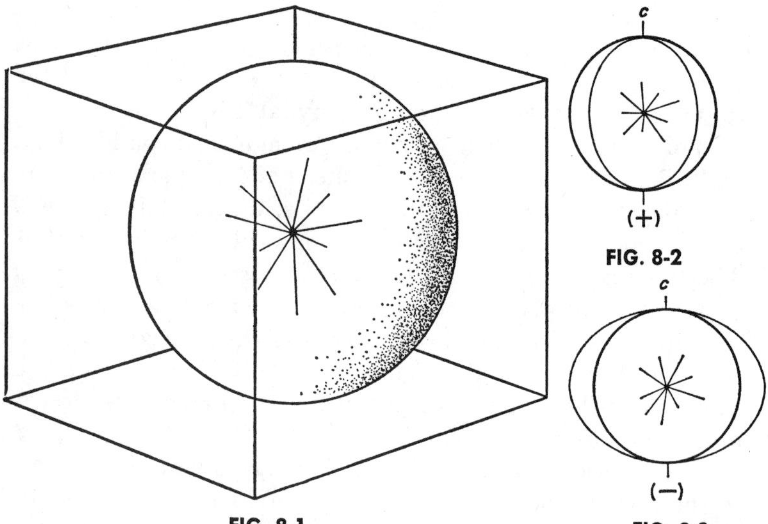

FIG. 8-1

FIG. 8-2

FIG. 8-3

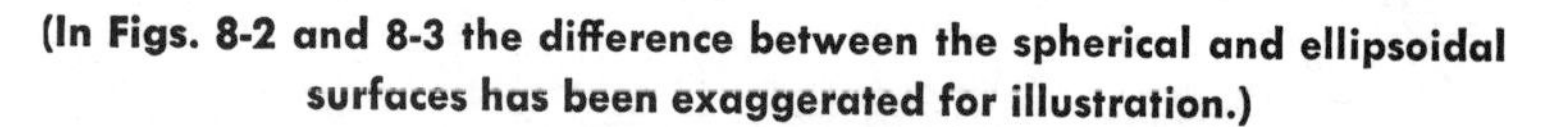

(In Figs. 8-2 and 8-3 the difference between the spherical and ellipsoidal surfaces has been exaggerated for illustration.)

have in glass, in air, or in a vacuum. The ray-velocity surface is a sphere: cubic crystals are *optically isotropic.* All other crystals are *optically anisotropic.* In trigonal, tetragonal, and hexagonal crystals the light traveling in a given direction is broken up into two rays, polarized at right angles to each other, traveling at different velocities. There are thus two velocity surfaces. One is spherical, like the surface in ordinary glass, ordinary air, and somewhat-less-ordinary cubic crystals, and is the ray-velocity surface of the *ordinary ray*. The other is an ellipsoid of revolution and is the ray-velocity surface of the *extraordinary ray*.

The spherical, ordinary ray-velocity surface coincides with the extraordinary ray-velocity surface at two points only. Consider the geometry of Figs. 7-5 and 7-6 relative to a sphere, and you will see where these points must be. They must be the ends of the axis of revolution of the ellipsoid, so that a cross section of the ordinary ray-velocity surface and the extraordinary ray-velocity surface looks like Fig. 8-2 or 8-3, according to whether the crystal is optically "positive" or optically "negative." The terms "positive" and "negative" are arbitrary. They could as well have been "A" and "B."

If the extraordinary ray is the *slow ray,* relative to the ordinary ray, then the ellipsoid will lie inside the sphere, as in Fig. 8-2, and the crystal will be said to be optically positive. If the extraordinary ray is the *fast ray,* relative to the ordinary ray, then the ellipsoid will lie outside of the sphere, and the crystal will be said to be optically negative. Note that the ray-velocity surface of the ordinary ray which is spherical is drawn with a heavier line than that of the extraordinary ray in these figures. The ordinary ray is of course the fast ray in positive crystals and the slow ray in negative crystals.

Light traveling along the axis of revolution of the ellipsoid (vertical in Figs. 8-2 and 8-3) will have a single velocity. This direction is called the *optic axis*. Only along this direction does light pass through a crystal as though it were glass.* In every other direction the light is broken up into two rays, both polarized, which travel with different velocities. The difference in velocity increases from zero along the optic axis to a maximum for propagation normal to the optic axis, as shown in Figs. 8-2 and 8-3. In general, the vibration direction of the polarized extraordinary ray is in the plane defined by the propagation direction (ray direction) and the optic axis. That of the ordinary ray is normal to this and normal to the propagation direction

* For rare exceptions to this statement, see Chapter 15.

of the ray. These directions are indicated by short dashes on the ray-velocity surfaces for the ordinary and extraordinary ray shown in Figs. 8-4 and 8-5. In the *a* figure of each, the optic axis is

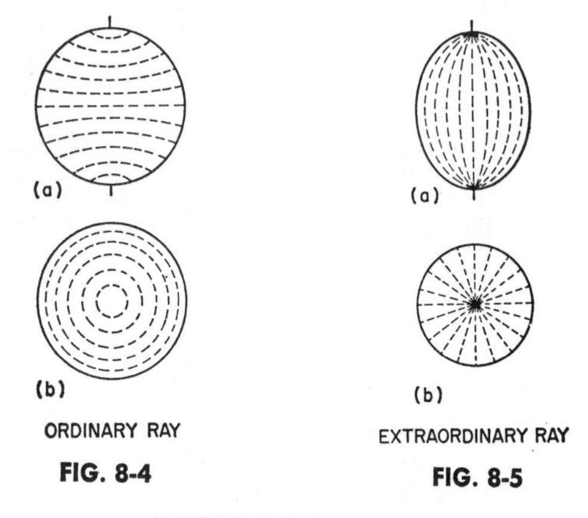

FIG. 8-4 **FIG. 8-5**

VIBRATION DIRECTIONS

lying in the plane of the paper; in the *b* figure, the optic axis is normal to the paper. (The vibration direction of the extraordinary ray is almost, but not quite, normal to its propagation direction. This is discussed in detail in the books by Wahlstrom and by Hartshorne and Stuart, listed at the end of this book.)

Notice that, no matter what the direction of travel from the center, the vibration direction of the ordinary ray will always be normal to that of the extraordinary ray.

Symmetry considerations show, as in the thermal-expansion case, that such ray-velocity figures are not compatible with cubic point-group symmetry but are compatible with all other point groups. Again, as in the thermal-expansion case, since they are not required by the symmetry of the orthorhombic, monoclinic, and triclinic crystals, they are in fact only exhibited by the tetragonal, trigonal, and hexagonal crystals. Since these ray-velocity figures have one optic axis, the tetragonal, trigonal, and hexagonal crystals are said to be *optically uniaxial;* the optic axis will, by symmetry, coincide with the *c* axis in each case (the 4-fold, 3-fold, or 6-fold axis, respectively). The orthorhombic, monoclinic, and triclinic crystals which have ray-velocity surfaces with two optic axes are said to be *optically biaxial.* Biaxial crystals will be discussed in Chapter 13. Unlike the cubic crystals, all of these are of course optically anisotropic.

In all optically anisotropic crystals, light traveling in any direction except along an optic axis is broken up into two rays with different velocities, vibrating normal to each other. In order to demonstrate the very beautiful results of the disparity in velocities, we need to be able to polarize the light before it goes into the crystal. A small sheet of polarizing film may be used. Such a sheet will polarize the light, letting through only that light which is vibrating in one particular direction.

A simple test will show what direction in the sheet this is. Light coming to your eye from any brightly reflecting nonmetallic surface such as the surface of still water, a flat piece of glass, or even an enameled or polished window sill will be mostly polarized: its vibration direction will be normal to the propagation direction of the light (a line from the surface to your eye) and parallel to the reflecting surface. (See Figs. 8-6 and 8-7.)

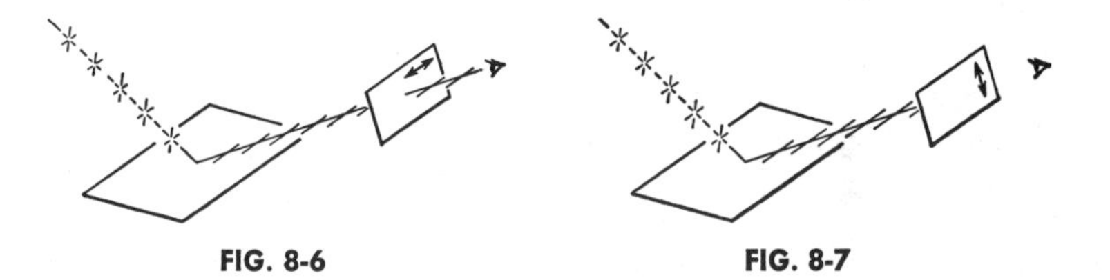

FIG. 8-6 FIG. 8-7

If the polarizer is so oriented that its own vibration direction is parallel to this vibration direction, it will let the polarized light through to your eye as in Fig. 8-6. If it is normal to this direction (Fig. 8-7), it will not.

For convenient reference, mark the vibration direction on the polarizing sheet. Although it is shown parallel to the edge of the square in Fig. 8-6 and subsequent figures, you should not assume this to be so in the case of any particular piece of polarizing film without testing it.

If a piece of polarizing film is placed on top of another with the polarization of the first normal to that of the second, as in Fig. 8-8, no light from below will come through both of them.

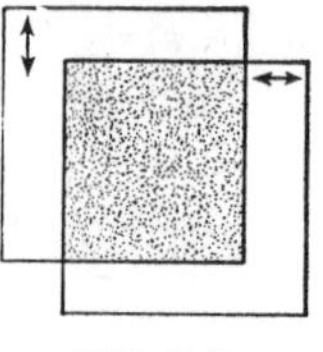

FIG. 8-8

(Actually, a little comes through because the polarizers do not polarize the light completely.) Since light with the only vibration direction that the upper polarizer lets through has already been excluded by the lower polarizer, there is none of it left for transmission by the upper polarizer. Similarly, light with the only vibration direction transmitted by the lower polarizer is not passed by the upper polarizer.

Now let us consider a slice cut from the tetragonal crystal, rutile (Fig. 5-1), parallel to the *c* axis—i.e., parallel to the optic axis. Light coming through it will be traveling normal to the optic axis. In terms of Figs. 8-4a and 8-5a, this light travels from the center of the solid vector figure toward the reader along a direction normal to the paper. In the crystal it will be broken up into the extraordinary ray, vibrating parallel to the optic axis, and the ordinary ray, vibrating normal to the optic axis. These vibration directions are shown diagrammatically in Fig. 8-9.

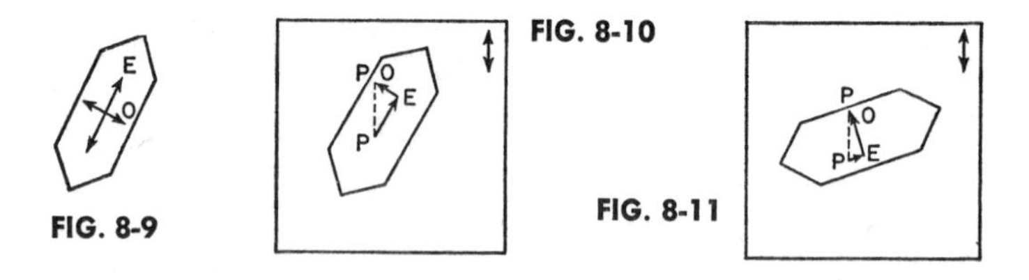

FIG. 8-9

FIG. 8-10

FIG. 8-11

Now if this crystal is placed on a piece of polarizing film as shown in Fig. 8-10 so that the light entering it from below is all vibrating in the direction *P-P*, it will analyze this vibration into the two components of motion permitted in the crystal, the E vibration direction and the O vibration direction. From a comparison of Figs. 8-10 and 8-11, you can see that the proportion of the incoming light that becomes the E ray relative to the proportion that becomes the O ray will depend on the orientation of the crystal relative to the polarizer. If the crystal were turned still farther so that the E vibration direction was normal to *P-P*, then the E component would be zero, and all the light would be transmitted as the ordinary ray, vibrating parallel to *P-P* (Fig. 8-12).

Now with the crystal again turned away from the position shown in Fig. 8-12, suppose the second polarizer is placed on top of the crystal with its polarization direction normal to the first polarizer. The crystal looks bright between the two polarizers. In other words, light now comes through the upper polarizer where the crystal lies between them (Fig. 8-13).

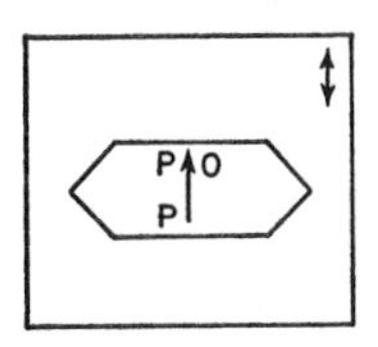

FIG. 8-12

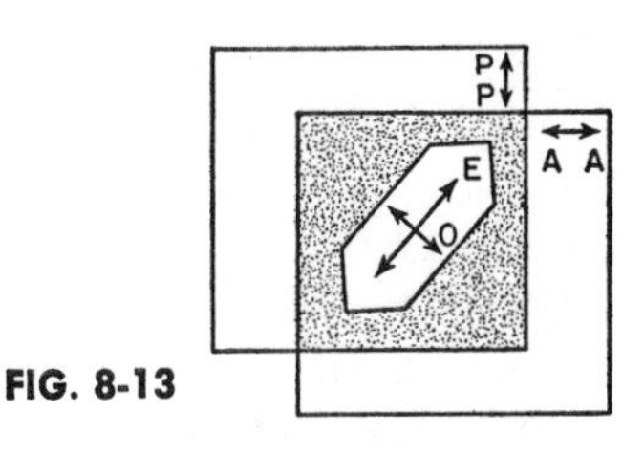

FIG. 8-13

If we again think of analyzing the vibration direction into two component vibration directions at the upper polarizer, we will see why the light comes through. Both the E and the O vibration direction at the upper polarizer of Fig. 8-13 may be thought of as having a component parallel to *P-P* and a component parallel to *A-A*, as in Fig. 8-14. The *P-P* components are

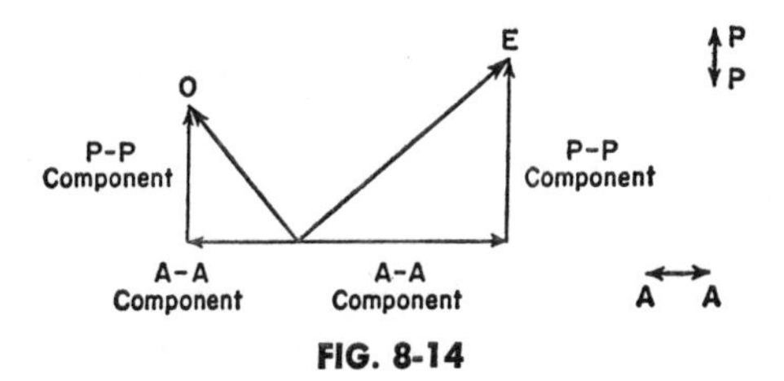

FIG. 8-14

not allowed to pass through the upper polarizer, but the *A-A* components are, and therefore light does get through, though it is less than what we started with because we had to leave part of it behind at the upper polarizer. The first polarizer in such a system is called the *Polarizer,* and the second, the one between the crystal and the eye, is usually called the *Analyzer.* In nearly all crystal work they are used with the polarization of one normal to that of the other (usually with the Polarizer up and down the page, or "north-south," front to back in a microscope, and the Analyzer left to right). The crystal is then said to be viewed in "crossed polarized light" or "between crossed polarizers." The former phrase is not clearly descriptive and therefore the latter phrase is preferred, but both are widely used.

When the crystal is rotated to the position shown in Fig. 8-12, then all the light coming through the crystal is in the *P-P* direction. An analysis like that in Fig. 8-14 would give a zero length for the *A-A* component, inasmuch as no light would come through the analyzer. Such a position is called an *extinction position.*

A direction in the crystal which is parallel to the polarization direction of the Polarizer or Analyzer when the crystal is at an extinction position is called an *extinction direction.*

How many times will the crystal be in an extinction position during a complete rotation of the crystal through 360° around the "line of sight" between crossed polarizers? Since it will happen every time the vibration directions in the crystal are parallel to the vibration directions of the Polarizer and Analyzer, it will happen every 90°—i.e., four times during a 360° rotation. *Note that this statement is true for every bit or fragment or slice or bar of every optically anisotropic crystal, regardless of its symmetry, except when it is viewed along an optic axis.*

What does happen when a crystal is viewed between crossed polarizers along the optic axis? Suppose that, from the tetragonal rutile crystal of Fig. 5-1, we cut thin slices normal to the *c* axis and observe them between crossed polarizers (Fig. 8-15). We see that no matter how they are rotated around the *c* axis they remain dark between crossed polarizers. The reason is that light traveling along the optic axis is not broken up into two rays vibrating normal to each other. It travels through the crystal as it would through glass.

Those who have looked at crystals between crossed polarizers will be protesting that we have omitted the most striking aspect of their appearance, their color. A colorless anisotropic crystal between crossed polarizers may show bright colors. This is caused by the fact that the two rays traveling in a given direction do not have the same velocity, and therefore one gradually gets ahead of the other. The amount that one is ahead of the other will depend on the difference in their velocities and the distance they have traveled with these different velocities, just as it would with two automobiles. When this amount is small (s_1, Fig. 8-16),

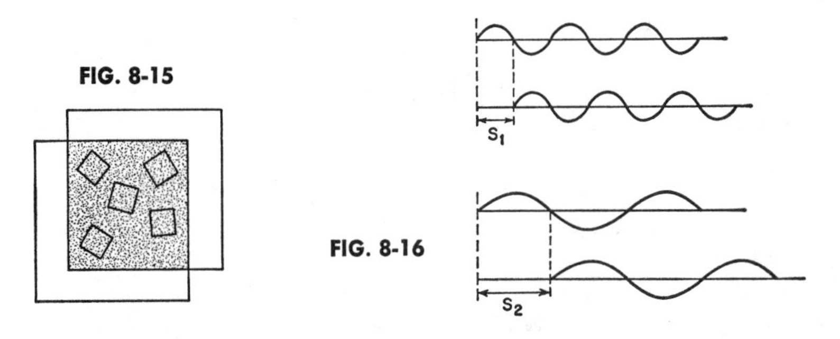

FIG. 8-15

FIG. 8-16

light with some particular short wavelength in one ray will be just out of phase with light with the same short wavelength in the other ray. When we look at those components of the light that are all vibrating in the same plane (selected by the Analyzer, the upper polarizer), we find that this short wavelength has been subtracted from the white light, which therefore shows the "complementary color," the color that is left when one of the constituents of white light has been subtracted. Such a color, which is due to the interference of one ray with another that lags behind it, is called an *interference color.* If the lag is greater, as s_2, Fig. 8-16, the wavelength that is just out of phase is a longer wavelength. This means that for a given slice of an anisotropic crystal between crossed polarizers, the color will vary with the thickness of the crystal. Such variation is shown in

Plate IV(1), which shows a piece of mica between crossed polarizers. Because of the easy cleavage of mica, bits can be peeled off, leaving different thicknesses in different parts of the mica. The square grid of lines shows the vibration directions of the Polarizer and Analyzer.

The short wavelengths of the visible spectrum give us the color violet. When these are subtracted, the resulting color is yellow. Somewhat longer wavelengths are blue. When these are subtracted, we have orange. Next, the subtraction of green gives us red, and then yellow, orange and red, subtracted in turn, give us violet, blue, and green. Then we start over again as the violet light of one ray now lags behind the other by exactly 1½ wavelengths. These successive sequences of interference colors are called successive *orders* of interference colors, numbered 1st, 2nd, 3rd, etc., as shown on Plate IV(2), which shows the interference colors from a wedge-shaped piece of quartz viewed normal to the optic axis between crossed polarizers.

Where the quartz crystal is thicker, two rays of one color are cancelled from the beam reaching the observer due to a 2½ wavelength lag while two of another color are cancelled due to a 3½ or 4½ wave-length lag. We therefore no longer have the simple subtractive colors of the first and second order, and the succession of higher orders is marked by an alternation of pink and green, getting paler with higher orders. Thus a crystal that just looks white between crossed polarizers is usually showing very high-order interference "colors," as, for example, the fragments of calcite that are not at an extinction position in Plate IV(5).

Very thin crystals may look white because the retardation is less than enough to show first-order yellow. See Plate IV(1). Only when the retardation of one ray behind the other approaches half the wavelength of visible violet light do we begin to see the subtractive color yellow. When the retardation is less than this, but not zero, light of all visible wavelengths is still coming through, giving us white light.

The maximum velocity-difference between the O and E ray in quartz is much less than it is in calcite. Therefore in Plate IV(5), the fragments of quartz all show lower interference colors than those of the calcite. One which is thick in the middle and thin on the edges is blue in the middle with the decreasing interference-color sequence (red, yellow, white) outward to the thin edge, as in the case of the mica in Plate IV(1). Other grains of quartz in Plate IV(5) that show only the lower interference colors may be thinner or may happen to be oriented so that the line of sight makes a smaller angle with the optic axis, in which case the velocity difference will be less. Some of the quartz grains are at extinction.

From Figs. 8-2 and 8-3 we recall that the difference in velocity between the two rays is zero for propagation in the direction of the optic axis and increases to a maximum for propagation normal to the optic axis. Because of this variation, the interference color exhibited by a crystal between crossed polarizers will depend not only on its thickness, but also on the direction in which the light is traveling through the crystal. In a thin slice of a rock, differently oriented grains of the same mineral will show different interference colors. If you rotate a crystal between crossed polarizers in such a way that the direction of the light path through the crystal changes, the interference color will change.

Twinning in crystals can be detected between crossed polarizers in some cases and not in others. If a trigonal crystal is twinned so that the two parts are related by a rotation of 180° around the 3-fold axis, the optic axis in one part is parallel to the optic axis in the other and the two orientations are indistinguishable between crossed polarizers. If, however, the optic axis is differently oriented in the two parts of the twin, they may easily be distinguished from each other. A good example is shown in the two crystals of barium titanate ($BaTiO_3$) between crossed polarizers in Plate IV(4). These are tetragonal crystals and therefore optically uniaxial. We are looking through thin plates which lie in the plane of the paper. In part of each crystal, the optic axis (parallel to *c*) is normal to the plate, but in part it lies in the plate. You can easily tell which is which. Light traveling along the optic axis is not broken up into two rays with different velocities and therefore cannot exhibit interference colors.

To make sure that the black part of such a twin does not

just happen to be "at extinction" (as in Fig. 8-12), it should be rotated around the direction of propagation of the light. If the light is in fact traveling along the optic axis, the crystal will remain dark during this rotation.

A physicist had a very pure specimen of the trigonal crystal corundum (Al_2O_3). It had been grown in the laboratory and was transparent and colorless. Such a corundum crystal is now called "sapphire." He sent it to a crystal shop to have a cube cut out of it, specifying that one pair of faces of the cube should be normal to the *c* axis, since he wanted to measure some property in a direction parallel and normal to this axis. When the cube was delivered to him, there was no indication of which pair of faces was normal to the *c* axis.

Between crossed polarizers he quickly determined the orientation. With the light traveling normal to the desired pair of faces, the crystal remained uniformly dark throughout a complete rotation around the light path. All other faces, being parallel to the optic axis, had extinction positions which occurred whenever their edges were parallel to the vibration directions of the Polarizer and Analyzer.

Some substances that are not commonly considered crystalline are optically anisotropic. For example, most cellophane shows interference colors between crossed polarizers and of course is at extinction every 90° when rotated around the direction of the light path. Apparently in the course of its fabrication into thin sheets, the molecules of which it is composed get sufficiently oriented so that in the sheet light is broken up into two rays, traveling with different velocities and with mutually perpendicular vibration directions. Is it a crystal? The author's opinion is that although it shows some orderliness it is not quite orderly enough to be included as a crystal. But some people might disagree.

Even cubic crystals may show light between crossed polarizers if they are optically active or badly strained. In the latter case the extinction direction varies from one part of the crystal to another, and this "wavy extinction" is readily distinguished from the uniform extinction of a non-cubic crystal.

Interference colors can be seen between polarizers that are

not crossed. Referring to Fig. 8-14, we recall that the function of the Analyzer was to select those components of the two vibrations that lay in the same plane. The *A-A* components were selected. But suppose the *P-P* components had been selected, i.e., suppose the Analyzer had been placed with its polarization direction parallel to that of the Polarizer. Then what would we see? We would still see interference colors, but not the same ones. The many possible variations resulting from various relative positions of Polarizer, Analyzer, and crystal will not be explored here. They can be understood on the basis of the effects discussed here so that the reader may experiment for himself.

Interference colors can be seen under other conditions than the observation of anisotropic substances with a Polarizer and Analyzer. Whenever light is reflected from a very thin transparent film so that the rays reflected from the bottom surface of the film (*A* in Fig. 8-17) rejoin those being reflected from the

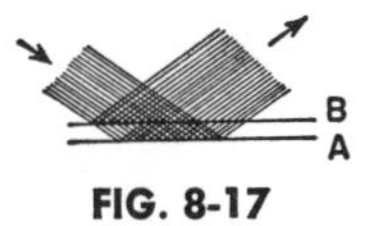

FIG. 8-17

top surface (*B* in Fig. 8-17) after traveling an extra distance, the *A*-reflected ray will be retarded relative to the *B*-reflected ray, and interference will take place. The thicker the film, the more the retardation will be. The same sequence of colors as that shown by the quartz wedge or by the mica in Plate IV can often be observed at the edge of a film of oil on the surface of water on the street.

If a drop of thin lubricating oil is put on the surface of cold water in a glass or cup, it does not spread out and stays too thick to show interference colors. However, if it is placed on the surface of very hot water, it spreads in a thin film and then produces interference colors which show up well if the observer looks toward the reflected light, preferably with his line of sight at a low angle to the surface. The colors change as the oil "pulls itself together" into smaller areas because of surface tension and thus becomes thicker as it cools.

The colors of soap bubbles are of the same origin, as are those of many birds, including the peacock and the hummingbird. A delightful paperback book about soap bubbles, written especially for young people by C. V. Boys, discusses their colors. The title is included in the list at the end of this book.

Robert Hooke, whose insatiable curiosity caused him to study clocks, the solar system, respiration, weather forecasting, and soap bubbles, among other things, discussed the colors of soap bubbles under the name of "fantastical colours" in his *Micrographia,* published in 1665. It was Sir Isaac Newton, however, who analyzed carefully the manner in which interference colors are produced when two slightly curved pieces of glass are placed close together. He included a discussion of this phenomenon in his "Discourse on Light and Colours" in his *Opticks,* book ii, published in 1675. To this day, interference colors produced in this way are known as Newton's rings. They may often be seen from two microscope slides that are stuck together or tightly pressed together. These are best observed against a dark background with a diffuse light source, such as light from the sky.

* * *

1. In Fig. 8-15, how could you determine optically that the crystals were not cubic crystals?
2. Calcite cleavage fragments nearly always lie on a cleavage face. As they lie on a glass microscope slide on the stage of the polarizing microscope, how will their extinction directions be related to the directions of the edges of the cleavage face?

* * *

Answers to questions at the end of Chapter 7

To determine the vector surface for the linear thermal expansion of a crystal:

1. In a tetragonal crystal, the surface is an ellipsoid of revolution (Fig. 7-5 or 7-6) with its axis of revolution along the *c* axis of the crystal and its circular equatorial section normal to this. Therefore two measurements will suffice, one to give us the length of the axial vector and the other to give us the length of the vector that is the radius of the circular equatorial section. These two will define the figure: thermal expansion measurement of a rod cut with its length parallel and thermal expansion measurement of a rod cut normal to *c* (Fig. 8-18).

2. In an orthorhombic crystal, the surface is a triaxial ellipsoid (Fig. 7-7) with its three axes lying along the three crystallographic axes, in accordance with the symmetry. A knowledge of the lengths of the three axes of such a figure suffices to define the figure. Therefore we need to measure the thermal expansion of three rods, one cut parallel to *a*, one cut parallel to *b*, and one cut parallel to *c* (Fig. 8-19).

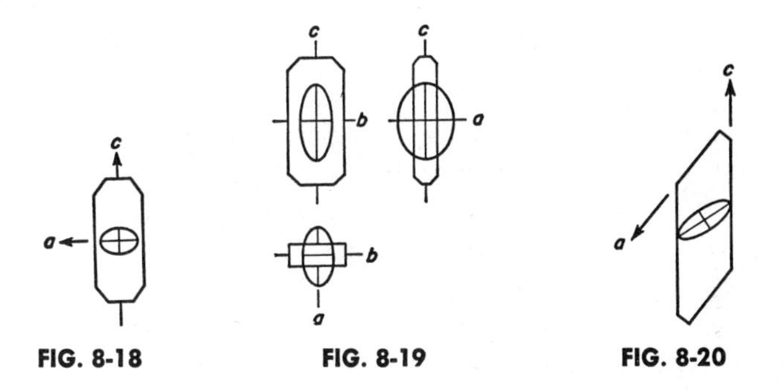

FIG. 8-18 FIG. 8-19 FIG. 8-20

3. In a monoclinic crystal, the surface is a triaxial ellipsoid. Symmetry requires that one of the axes of the ellipsoid lie along the *b* axis of the monoclinic crystal since this is either a 2-fold axis or the normal to a symmetry plane, according to the point group of the crystal. Therefore we measure the expansion along *b* in one rod whose length lies along *b*. The other two axes of the triaxial ellipsoid may lie anywhere in the plane normal to *b* (the plane containing the *a* and *c* axes, which is therefore commonly called the *ac* plane). So we are faced with the problem of determining not only the dimensions but also the orientation of the ellipse which is the cross section of the expansion surface in this plane (Fig. 8-20).

If it were a circle instead of an ellipse, one measurement would suffice (just as it does for the spherical vector surface of the cubic crystal) because there is only one unknown, the radius. In the case of the ellipse in the *ac* plane, there are three unknowns: the major axis of the ellipse, the minor axis of the ellipse, and the angle between an ellipse axis and a crystallographic axis (either *a* or *c*). Therefore we need three measurements so that we can set up three equations to solve for the three unknowns.

9 *Uniaxial Crystals in Convergent Polarized Light*

In Chapter 8 we considered the optical effects observed when a crystal is viewed in crossed polarized light. When the light is traveling to the eye along the optic axis (Fig. 9-1), the crystal looks dark between crossed polarizers, but at other angles interference colors are seen except when the crystal is at extinction—i.e., with its vibration directions lying in the planes of polarization of the Polarizer and Analyzer.

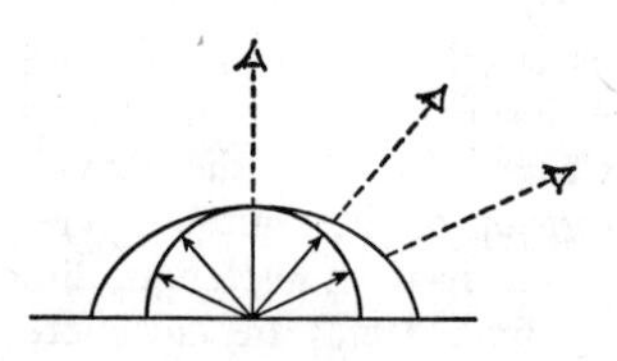

FIG. 9-1

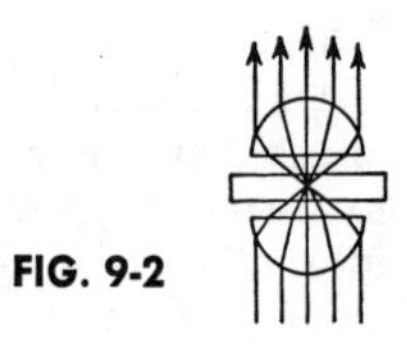

FIG. 9-2

Very instructive and beautiful effects can be observed by looking through the crystal in more than one direction at once. This is not so impossible as it would at first seem. With a suitable lens to focus the light so that it passes through the crystal in many directions and another lens to catch the rays that have so passed and direct them along the barrel of a microscope (Fig. 9-2), one may indeed look along all those directions contained within a cone whose angle will depend on the way the lens is made.

If we do this while looking along the optic axis, with crossed polarizers placed anywhere in the light path (so long as one of them comes before the crystal and the other after the crystal) and with the ocular of the microscope removed,* the splendid figure shown in Plate V(1a and 1b) is seen. This, of course, is

* An alternative to removing the ocular of the microscope will be discussed in Chapter 10.

just what you expected from the properties of uniaxial crystals discussed in Chapter 8. The center is black because here the light is traveling right along the optic axis. In all directions outward from the center the light path through the crystal has made an angle with the optic axis which is small near the center of the figure and larger farther out. Therefore, as you would expect from Fig. 9-1, the retardation of one ray behind the other is small near the center of the figure and larger farther out so that, from the center outward, we see the increasing sequence, yellow, red, blue, green, and so on up to higher orders in which the colors simply alternate between pink and green for the reason discussed in Chapter 8. The retardation in calcite, shown in Plate V(1b) is greater than that in quartz, shown in Plate V(1a), and therefore the interference colors increased more rapidly, from the center outward, in Fig. 1b.

Such a figure, which can only be observed in convergent (and divergent) light between crossed polarizers and shows the sequence of interference colors resulting from the different directions of the light path through the crystal, is called an *interference figure.*

We now understand the dark center and the increasing order of interference colors giving the concentric color bands of the interference figure. What about the black cross? In Figs. 8-4 and 8-5 the short dashes showed the vibration directions of the ordinary and extraordinary rays as they progressed outward in all directions from an imaginary point-light-source in the center of the crystal.

Suppose we look down on the "north pole" of such a figure as we are doing in Figs. 8-4b and 8-5b. Combining these two figures, we get Fig. 9-3, which shows us the pairs of vibration directions of light contributing to various points of the uniaxial interference figure. (The vibration directions of the extraordinary ray are indicated by heavy dashes, those of the ordinary ray by light dashes.) Since we are seeing the interference figure with the crystal between crossed polarizers, the polarization directions of these are marked in Fig 9-3 (*P-P* for Polarizer; *A-A* for Analyzer). The shaded part of the figure shows where the vibration directions of the rays lie parallel or nearly parallel to the planes of polarization of the Polarizer and Analyzer. Of course no light or very little light can come through in this part, and so we get the black cross.

The optic axis (normal to the paper in Fig. 9-3) is an axis of

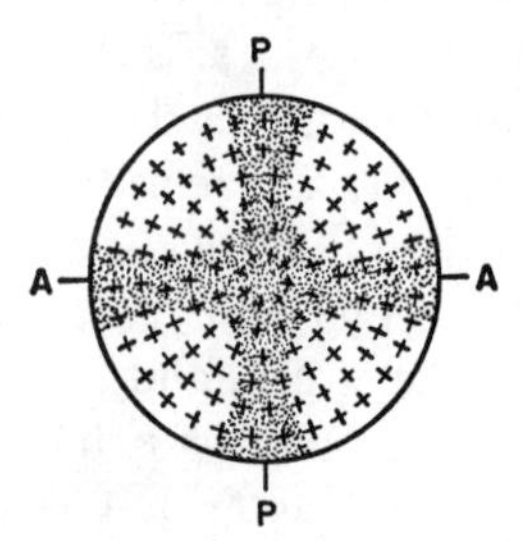

FIG. 9-3 Heavy radial dashes indicate the vibration directions of the extraordinary ray; light dashes, those of the ordinary ray.

infinite optical symmetry, and therefore rotation of the crystal around this axis between crossed polarizers causes no change in the interference figure.

Suppose we translate the crystal plate parallel to itself, just sliding it from left to right, for example, on the stage of the microscope. Again, if the crystal is uniform in thickness, there is no change in the interference figure because it, like the other properties of a crystal, is the result of the orderly arrangement of atoms in the crystal; if we are looking along the *c* axis of this arrangement, it makes no difference which particular group of unit cells we are looking through. They are all alike.

If, however, we tilt the crystal relative to our line of sight so that the light comes through the crystal along a path which is not parallel to *c*, then the appearance of the figure will change.

If the path of the light through the crystal to the eye makes a small angle with the optic axis, then, as shown in Fig. 9-4a, if the crystal is rotated around this path direction, the axis will rotate around the path direction, moving on the surface of an imaginary cone. The center of the cross will of course move in a circular path in the field of view, as seen in Figs. 9-4b to e, where the arrow outside the circle shows the direction of rotation of the crystal. The arms of the cross will remain parallel to the polarization directions of the crossed polarizers.

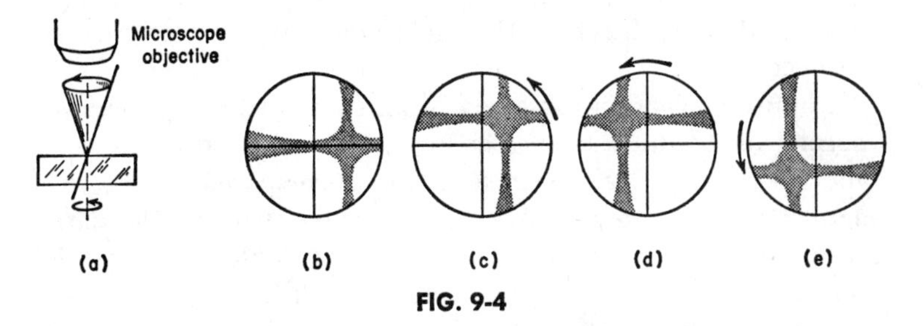

FIG. 9-4

The interference figure that is seen when a uniaxial crystal is viewed in convergent light between crossed polarizers with its optic axis normal to the line of sight (= light path) is called the *flash figure* for a reason that will be clear from Fig. 9-5. As in the

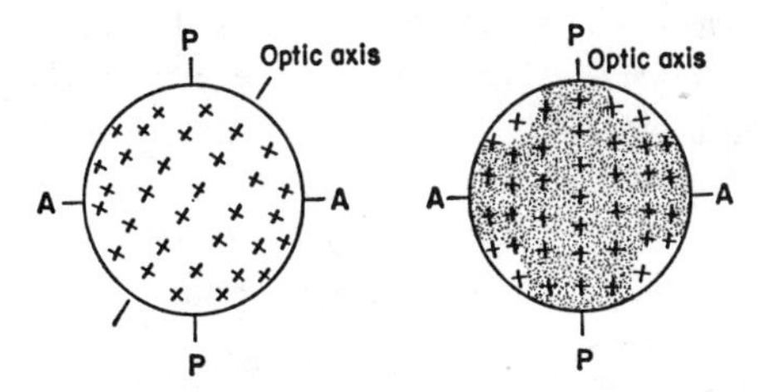

FIG. 9-5 Heavy dashes parallel or nearly parallel to the optic axis indicate the vibration directions of the extraordinary ray; light dashes, those of the ordinary ray.

case of Fig. 9-3, this figure shows the vibration directions of the ordinary and extraordinary rays and is shaded wherever these vibration directions are parallel or nearly parallel to the planes of polarization of the Polarizer and Analyzer.

As the crystal is rotated around the line of sight (normal to the paper in Fig. 9-5), the angle between the crystal vibration directions and the polarization planes of the Polarizer and Analyzer changes. Only when the optic axis lies almost parallel to the plane of polarization of either the Polarizer or the Analyzer is much of the area of the interference figure dark. Then nearly all of it is dark at once. It is because this happens for such a small angular rotation of the crystal that this figure is called the flash figure. Compare this case with that of Fig. 9-3, where the vibration directions remain unchanged as the crystal is rotated and therefore the interference figure remains unchanged.

The colors that are observed in an interference figure viewed normal to the optic axis, as in Fig. 9-5, are generally pale pinks and greens in poorly defined patterns, compared to the axial figure. Referring to Figs. 8-2 and 8-3, we see that the retardation is highest for this direction, and therefore, unless the crystal is very thin the colors are apt to be high-order colors. The velocity difference between the two rays in this direction changes less rapidly with change of angle than it does near the direction of the optic axis, so that the change of interference colors is more gradual and the color bands are therefore broad.

Under favorable circumstances it is possible to see an interference figure without the aid of lenses. If the eye is placed very close to the crystal, then the light entering the eye has come through the crystal from a range of directions (Fig. 9-6) and an

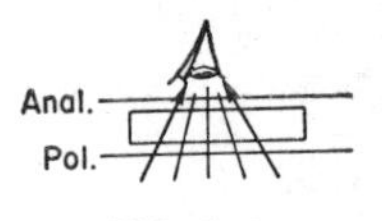

FIG. 9-6

interference figure can often be observed, especially if one is looking along an optic axis so that the figure is well defined and colorful.

In Plate V(3) we see a broken piece of a hexagonal crystal (guanidinium aluminum sulphate hexahydrate) showing an interference figure in this way. The large surfaces of the plate are {0001} cleavage planes, so the optic axis is normal to these. However, the plate is slightly tilted relative to the "line of sight" of the camera, and therefore the interference figure is a bit off center, as in Fig. 9-4.

This crude way of viewing an interference figure commonly does not give as good a figure as can be seen with the polarizing microscope, but for a quick approximate determination of the orientation of a large specimen it can be very useful. An example of its use in finding the *c* axis of a big beryl crystal (hexagonal) will illustrate a difficult case. The specimen was about

2 × 1.5 × 1 inches with all surfaces irregularly fractured, but it was clear and colorless. Reflection and refraction* of light at its surfaces made it impossible to get a good optic figure or even good extinction between crossed polarizers. However, when the crystal was held in a particular orientation, the bright interference colors of the first and second orders could be seen, unlike the whitish light observed for other orientations. For this orientation the retardation must have been low, i.e., the light must have been traveling nearly along the optic axis. Trial surfaces were cut on the specimen and the orientation confirmed.

In Chapter 1, the use of calcite in an optical ring sight or target finder was mentioned. A picture of the ring sight is shown in Plate VII(1), and a view through the sight is shown in black and white in Plate VII(2). If it were in color, we would see the familiar bands of increasing interference colors of a uniaxial interference figure. In the ring sight is a thin plate of calcite cut normal to the optic axis. The sight must be mounted with the optic axis of the calcite crystal parallel to the gun barrel so that when the target appears in the center of the interference figure, as in Plate VII(2), the gun will be aimed at the target.

The calcite plate must of course be mounted between crossed polarizers. How have the designers managed to get rid of the black cross and to let light from the target come through the center of the figure? The explanation involves a knowledge of circularly polarized light. Since this knowledge will not be needed for an understanding of the rest of this book, an explanation has been placed at the end of the book as Appendix I.

* * *

1. Sketch a sequence of diagrams like those in Fig. 9-4, but for the case where the center of the cross lies just outside the field of view because the optic axis is making a larger angle with the light path through the crystal.

Notice the motion of the arms of the cross in the field of view. This would be what you would see in the microscope, and from it you would have to deduce the position of the optic axis.

* See Chapter 12.

2. What would a uniaxial interference figure, like that in Plate V(1a), look like in sodium light? Why?

* * *

Answers to questions at the end of Chapter 8

1. If the crystals in Fig. 8-15 were cubic, they would remain dark in crossed polarized light in any orientation: tilted, on edge, or any other way. If the crystals in Fig. 8-15 do not do this, they are not cubic. Therefore you would tilt them and change the orientation of the crystal plate relative to the light beam. If they let light through in any orientation between crossed polarizers, they are not cubic. (For a rare exception to this statement, see Chapter 15.)

2. The rhombohedral shape of the solid in Fig. 3-8 is approximately that of the cleavage rhombohedron (Plate VI(2)). The optic axis must be parallel to the $\bar{3}$ axis in calcite, which is shown as the hexagonal *c* axis in Fig. 3-8.

In a beam traveling normal to the top front face of Fig. 3-8, for example, the ordinary ray will be vibrating normal to the direction of propagation and normal to the optic axis. This direction of vibration is parallel to a_2(H) in Fig. 3-8. This direction bisects the acute angles between the edges of the rhombohedral face. The vibration direction of the E ray, which must be normal to that of the O ray, bisects the obtuse angles between the edges of the rhombohedral face (Fig. 9-7).

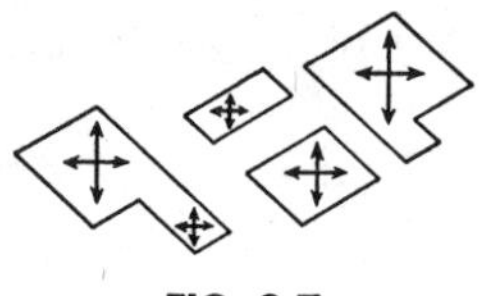

FIG. 9-7

Again, since we are concerned with *direction* in relation to the structure, the *orientation* of the extinction directions as given above will be unaffected by the physical dimensions (length, width and thickness) that the fragment happens to have (Fig. 9-7).

10 *The Polarizing Microscope*

The polarizing microscope is a microscope especially made for viewing objects in crossed polarized light. Like other microscopes, polarizing microscopes come in a wide range of types and prices, but nearly all of them have the essential components shown in Fig. 10-1.

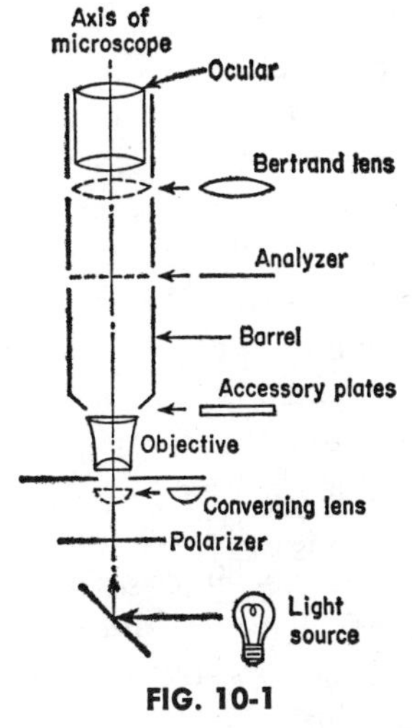

FIG. 10-1

The ocular ("eye-piece") and objective lenses serve the purpose of magnifying the object, as in other microscopes. A pair of cross hairs is attached to the ocular. These cross hairs are useful for reference in the field of view. (See, for example, Fig. 9-4.) Their orientation can be changed by rotating the ocular in the barrel, but they are most conveniently left in position parallel to the polarization directions of the Polarizer and Analyzer. In most microscopes there is a notch and pin arrangement which, when engaged, ensures the correct orientation of the cross hairs.

The stage on which the object is placed has a hole in it so

that the object may be viewed by transmitted light. It can be rotated in its own plane around the axis of the microscope so that the vibration directions of light in the crystal under examination can be placed at any desired angle to the planes of polarization of the Polarizer and Analyzer.

The Polarizer is held below the stage. It is generally left in place all the time, customarily with its polarization direction lying in the front-to-back (north-south) plane of the microscope, but in most microscopes it can be removed or rotated if desired.

The Analyzer is held in the barrel of the microscope, usually somewhere up near the middle, on a sliding carriage so that it can be slid in and out of the barrel at will. In some microscopes when it is in place in the barrel it can be rotated in its own plane around the axis of the microscope so that it can be "crossed" or "uncrossed" relative to the substage polarizer.

For viewing interference figures, convergent light is needed, as in Fig. 9-2. An auxiliary converging lens is therefore provided, which can be swung into position close beneath the crystal when needed. At the same time a highly magnifying objective is used above the stage. If the ocular is left in, then the Bertrand lens must also be inserted into the barrel of the microscope for viewing the interference figure. Like the Analyzer, this is mounted on a sliding carriage and can be slid in and out at will. If the Bertrand lens is not used, the interference figure may be seen with the ocular removed. It is not quite as easy to see in this way and appears smaller, but usually also sharper.

For many years the polarizing elements (Polarizer and Analyzer) in such microscopes have been Nicol prisms. Today some microscopes have Nicol prisms and some have a good grade of Polaroid polarizing film. A Nicol prism is a piece of calcite that has been cut apart and recemented in such a way that the ordinary ray is reflected off to one side when it meets the cemented surface, but the extraordinary ray passes through. The result is that the light emerging from the upper surface of the Nicol prism has the vibration direction of the extraordinary ray—i.e., it is polarized light.* The construction of the Nicol

* The Nicol prism as now used is constructed somewhat differently from the original one. William Nicol, who first thought of this ingenious way

prism is described in detail in the references at the end of this book, some of which also contain a much more detailed discussion of the polarizing microscope than is given here.

A slot in the barrel is provided for the insertion of accessory plates which will be discussed in the next chapter.

* * *

Answers to questions at the end of Chapter 9

1. Note that the center line of each arm of the cross always remains parallel to the polarization directions of the Polarizer and Analyzer. The motion of the arm across the field of view is translation without rotation. (See Fig. 10-2.)

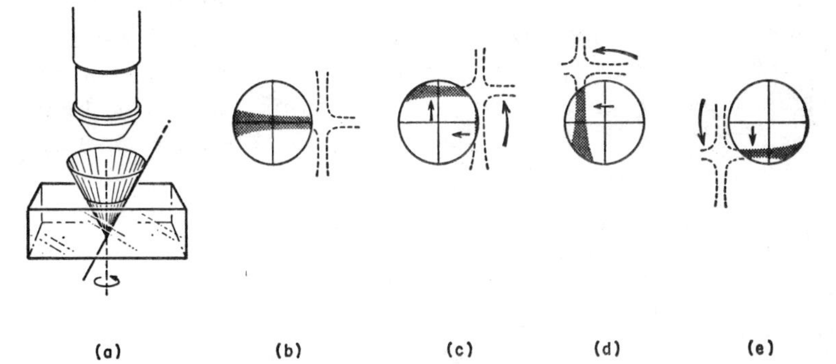

(a) (b) (c) (d) (e)

FIG. 10-2

2. A uniaxial figure in sodium light would show the usual black cross and a set of black rings on a yellow background instead of the colored rings. Only a narrow range of wavelengths, all yellow, is passing through the crystal. Where the retardation is just right for yellow light of the ordinary ray to be out of phase with yellow light of the extraordinary ray (retarded by $\frac{1}{2}$ wavelength, $\frac{3}{2}$, $\frac{5}{2}$, etc.) when it reaches the eye, a dark band will be seen.

---

of producing polarized light in 1828, was a Scottish physicist. During the early part of his life he gave popular lectures in science (then called "Natural Philosophy"), but apparently what he really wanted to do was work alone in his home laboratory with crystals and optical instruments since, when he had made enough money from fees charged for his lectures, he retired to live a very secluded life, studying crystals and other objects under the microscope. His chief interests were liquid-filled cavities in crystals and the cellular structure of fossil wood, but he was known to have been very skillful in grinding lenses, some of which he made from garnet and other semiprecious crystals.

11 *The Use of Accessory Plates: Determination of Optic Sign*

In Chapter 10 we saw that a polarizing microscope is equipped with an opening in its barrel for the insertion of "accessory plates." These plates make it possible to distinguish which of the two vibration directions of an anisotropic crystal belongs to the faster ray and which belongs to the slower ray.

The principle is very simple. All you need is a plate of an anisotropic crystal in which the vibration directions have been identified: one as the direction of the faster ray, the other as that of the slower ray.

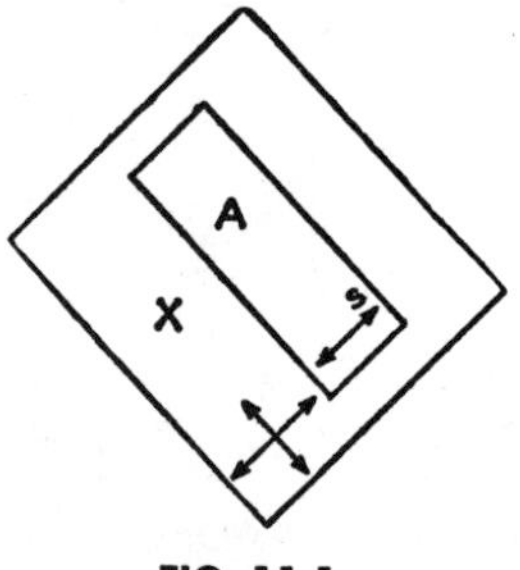

FIG. 11-1

Suppose this plate is given to us with the slow direction marked as on plate *A* in Fig. 11-1. Now we want to determine which of the vibration directions in crystal *X* is that of the fast ray and which is that of the slow ray. We lay the test plate, *A*, on top of the unknown, *X*, with their vibration directions parallel. This is easily done between crossed polarizers just by placing both of them at extinction. If we now rotate the pair together around our line of sight (normal to the page in Fig. 11-1) between crossed polarizers, they will become bright and we can examine their interference colors. Now where we are looking through both crystal plates, the colors we see will be due to the added effects of both crystals. If by chance we have placed the "slow vibration direction" of the test plate over the slow vibration direction of the unknown, the retardation in the test plate will increase the retardation that was begun in the unknown, and therefore the interference colors from the two together will be higher interference colors, just as they are from a thicker part of a single crystal plate. However, if by chance we have placed the slow vibration direction of the test plate over the fast vibration direction of the unknown, the retardation of the test plate will decrease the retardation that was begun in the unknown, and therefore the interference colors from the two together will be lower interference colors.

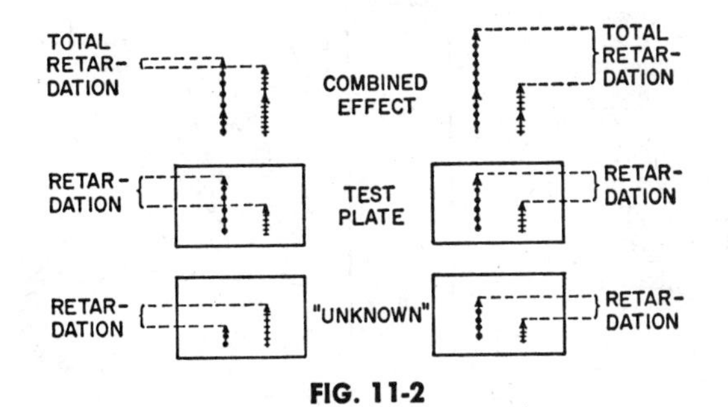

FIG. 11-2

Both cases are shown diagrammatically in Fig. 11-2, where we are looking normal to the light path. The lengths of the vectors represent the velocity of the light. The short dashes represent vibration direction in the plane of the paper: the dots represent vibration normal to the paper.

Plate I shows two sheets of mica, one partly on top of the other, between crossed polarizers. The sketch of this photograph given in Fig. 11-3 is for convenience in discussing it. The vibration directions of the two mica sheets are parallel to each other. (You could not tell this from the photograph, but they were tested as described above.) In Fig. 11-3 we are told which vibration direction in mica sheet number 2 is that of the slow ray and which is that of the fast ray. The problem is to determine the equivalent information for mica sheet number 1.*

Notice that although sheet number 1 has been cleaved irregularly so that it exhibits different thicknesses and therefore different interference colors (*a, b,* and *c,* Fig. 11-3) in different parts, the square sheet number 2 is uniform in thickness for the most part, giving a uniform interfence color (b in Fig. 11-3) over most of its surface except for a slightly thicker spot at the lower left which gives color *c.*

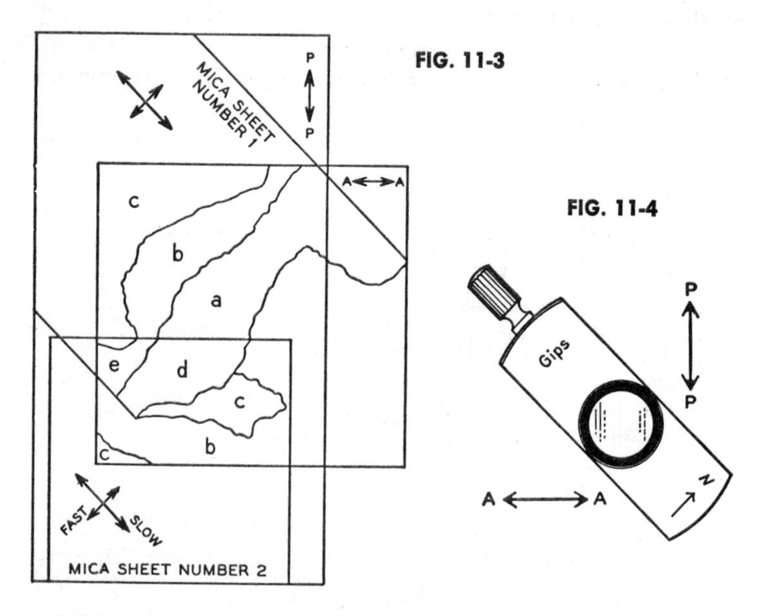

FIG. 11-3

FIG. 11-4

Where the two sheets are superposed, the interference colors are of course due to their combined effect. The reader is invited to determine for himself which is the vibration direction of the slow ray and which is that of the fast ray in sheet number 1. The answer will be given at the end of this chapter.

The accessory plates that come with the petrographic microscope are mounted in holders, as shown in Fig. 11-4, which are marked with the vibration direction of the slow ray. There is generally a gypsum plate (marked *Gips* in the German microscopes), a mica plate (*Glimmer*) with different amounts of re-

*Sheet number 1 is shown with small magnification in Plate IV (1).

tardation, and a quartz wedge, shown between crossed polarizers in Plate IV(2). The slot which admits these plates to the barrel of the microscope is so oriented that when the plates are in place their vibration directions are at 45° to the customary orientation of the polarization directions of the Polarizer and Analyzer. Since this orientation is fixed, one need only rotate the specimen until its vibration directions are also 45° to those of the polarizers (i.e., rotate it 45° from its extinction position) to align its vibration directions parallel to those of the accessory plate.

With certain interference colors, especially the higher-order ones, it is sometimes difficult to tell whether the color seen with the accessory plate in place is higher or lower than that without the plate. In such a situation the quartz wedge comes in handy. As the quartz wedge is inserted, the interference colors change. If the quartz retardation is adding to that of the "unknown," the colors will progressively rise as the thicker part of the wedge is inserted. If, on the other hand, the quartz retardation is subtracting from that of the "unknown," the colors will progressively fall. The specimen on the microscope stage can always be oriented so that the latter is the case, and it is a satisfying experience to see the confusing pale pinks and greens of the high orders give way to the vivid recognizable colors of the lower orders as the quartz wedge is inserted.

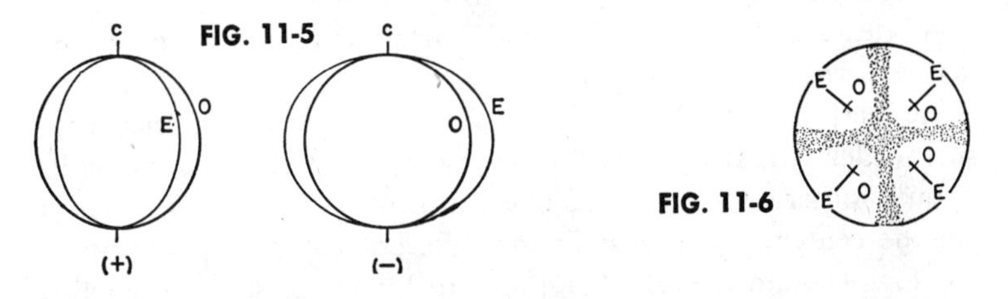

FIG. 11-5

FIG. 11-6

The observation of an interference figure with an accessory plate is very informative. Recalling, with the aid of Fig. 11-5, that there are two kinds of optically uniaxial crystals, "positive" and "negative," we see that in order to distinguish between the two possibilities we need to know whether the extraordinary ray is faster or slower than the ordinary ray. Recalling, with the aid of Fig. 11-6, the vibration directions of the ordinary and extraordinary rays in the uniaxial interference figure (shown more fully in Fig. 9-3), we see how beautifully the insertion of an accessory plate in the path of light giving such a figure will tell us what we need to know. Fig. 11-7 shows this diagram-

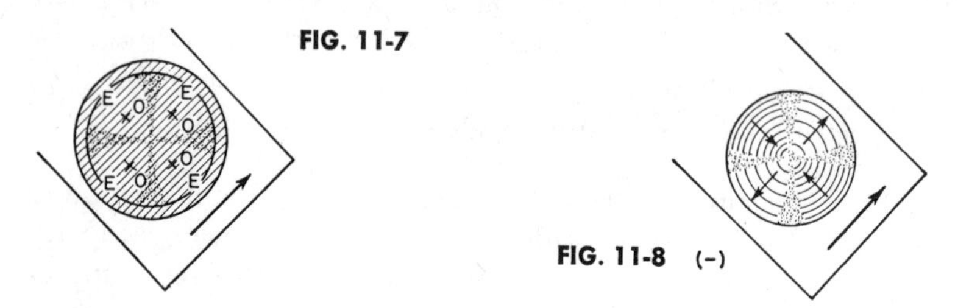

FIG. 11-7

FIG. 11-8 (–)

matically. The slow direction of the accessory plate is shown as a grid of fine lines crossing the interference figure. If the crystal is positive (E slow), then in the upper right and lower left quadrants we will have slow over slow, and the colors of all the bands will be higher than they were without the plate. In the upper left and lower right quadrants, however, we will have slow over fast (O fast), and the colors of all the bands will be lower than they were without the plate. Of course if the crystal is optically negative (E fast, O slow), it will be the upper left and lower right quadrants that show higher colors and the upper right and lower left that show lower colors.

Plate V(2), shows a uniaxial interference figure of quartz viewed with a gypsum plate in the conventional orientation. The plate, viewed by itself, would give the interference color first-order red. It therefore imparts this color to the cross, which would otherwise be black. Comparison of the colors adjacent to the center of the cross in the different pairs of quadrants, as in Fig. 11-7, to see whether they are higher or lower than first-order red, will show us the optic sign of the crystal photographed—i.e., whether it is positive or negative. Which is it? Reference to the interference-color sequence as shown by the quartz wedge in Plate IV(2) may be helpful. The answer is given at the end of this chapter.

If the quartz wedge is inserted in the accessory slot for testing an interference figure, the effect is spectacular. As the wedge is pushed in, the interference colors in two of the quadrants rise. In these quadrants the outer, higher-color, paler bands move inward continuously to occupy the positions of the inner, lower-color bands, which move before them to the center of the figure and disappear (Fig. 11-8). At the same time, the colors in the other two quadrants are falling. The inner, lower-color, brighter bands move outward continuously, following the paler bands to the rim of the field of view. This phenomenon must be seen to be appreciated!

* * *

Answer to question in Chapter 11 about Plate I: The vibration direction of mica sheet number 1 that is parallel to the vibration direction of the slow ray in mica sheet number 2 is the vibration direction of the slow ray. "Slow over slow" results in the higher interference colors where the two sheets are superposed.

* * *

Answer to question in Chapter 11 about Plate V(2): This is the interference figure of a positive uniaxial crystal, quartz.

The interference color of the gypsum plate alone is first-order red. The retardation in the gypsum plate is added to that of the quartz in the upper right and lower left quadrants where gypsum slow is parallel to quartz slow and gypsum fast is parallel to quartz fast. In the other two quadrants (upper left and lower right) the retardation in the gypsum plate is subtracted from that caused by the quartz plate. In these quadrants there is a black band where the retardation in the quartz is just exactly *compensated* by the retardation in the gypsum, so that the sum of the two retardations is zero.

* * *

EXPERIMENTS WITH TWO SMALL PIECES OF POLAROID POLARIZING FILM

Many hobby stores sell small squares of Polaroid film. With two of these you can perform an infinite number of experiments. If you have two pairs of Polaroid eyeglasses (so that you can place one over the other) they will serve nearly as well.

First determine the polarization direction of each piece of polarizing film as precisely as possible (Figs. 8-6 and 8-7) and indicate it by a short scratch in the corner of each piece. Then try the following experiments.

In the polarizing sheet are long molecules which are sufficiently well oriented parallel to each other to break up light into two rays, vibrating normal to each other, just as an anisotropic crystal does, and one of the two rays is very strongly absorbed. Therefore the light that gets through is polarized.

When the two sheets are superposed with their vibration directions parallel, light transmission is at a maximum. As one is rotated relative to the other, the intensity of the light diminishes with the square of the cosine of the angle of rotation, falling to zero when the angle is 90°. The intensity varies with the square of the vibration amplitude indicated by the vector component passed by the second sheet.

In all the following experiments the polarizers should be in the crossed position (minimum transmission) with the specimen between them. For convenience in handling, they may be clipped to opposite sides of a shallow box-lid.

Although cellophane is not a crystal, it contains long molecules, oriented in the fabrication of the sheet, and is optically anisotropic, so we will begin with cellophane because it is readily available and easy to handle and observe. The wrapper from a cigarette package is suitable.

1. Place the cellophane sheet between the polarizers with its surface parallel to theirs and looking through them toward the light. Rotate the cellophane around the line of sight and determine its extinction positions. Since the vibration directions of the two rays of light in the cellophane will be parallel to the polarization directions of the polarizers when the specimen is at extinction, these can now be marked on the cellophane. Later we will determine which is fast and which slow. (See Chapter 11.)

2. Cut rectangular pieces of cellophane with their edges parallel to the vibration directions. A single piece of cigarette-package cellophane commonly has a first-order pale gray interference color. If two are superposed, *maintaining their original orientation,* between crossed polarizers, the retardation is doubled and the interference color is yellow. (See Plate IV(2).) With three parallel sheets superposed the interference color is raised to first-order red, and with four parallel sheets superposed it is raised to blue (second order).

If one is rotated 90° around the line of sight so that the vibration direction of the fast ray in one becomes parallel to the vibration direction of the slow ray of the other, the retardation caused by the first piece is exactly undone or "compensated" by the second. (See Chapter 11.) The resulting effect is the same as though they were not there. A simple way to achieve this is by folding a cellophane sheet with the folded edge at 45° to the vibration directions.

3. A crystal which is repeatedly grown in every home today is ice. Crush an ice cube and observe some of the clear pieces between crossed polarizers. (*Caution:* Molten ice may damage the polarizers.) Notice

the effect of thickness on interference colors near the edge. Rotate the specimen around the line of sight. If one part is at extinction when another part is not, then the specimen consists of more than one crystal. All that part of the specimen that has a continuous, homogeneous structure (i.e., that is a single crystal) will be at extinction at the same time. Better specimens can be produced by freezing a drop of water on a microscope slide or other plane piece of glass. Since the glass is optically isotropic, it does not affect the interference colors. Boiled water makes clearer ice because the gas (which produces bubbles in the ice) is expelled during boiling.

The birefringence (see Chapter 12) of ice is low (.004), so that fairly thick pieces give good interference colors.

4. Salol (phenyl salicylate) melts at 43°C and can therefore be melted over a match flame on a glass slide. It is obtainable at a moderate price from most drug stores. Since the melt tends to supercool (not form crystals, even below the freezing point, 43°C), the addition of solid particles of salol may be required to cause crystallization. The crystal is orthorhombic, and its major faces are rhomb-shaped (like the "diamond" on playing cards). Because of high birefringence these crystals look white between crossed polarizers if they are allowed to grow thick. If they are grown between two pieces of glass that are held tightly together while the crystals are growing from the melt, bright first- and second-order interference colors may be seen. These colors are in striking contrast to the black that is seen when the isotropic molten material is viewed in crossed polarized light. Salol can be dissolved in alcohol and crystallized from the solution as the alcohol evaporates.

5. Sheets of mica are available from some scientific supply companies. The investigation of mica between crossed polarizers is discussed in Chapters 8 and 13. With the mica sheet parallel to the polarizers and at about 45° to extinction, the various parts of the biaxial interference figure may be seen if the mica (between polarizers) is held close to the eye and tilted around the y axis. (See Chapter 12.)

6. Accessory plate. (See Chapter 11.) In all pieces of "Scotch" tape tested by the author, the long direction of the tape is the vibration direction of the slow ray. If this is always true, then the vibration directions of the fast and slow rays in other substances can be determined by using a piece of Scotch tape as an accessory plate as described in Chapter 11. It is more conveniently handled if stuck to a glass slide. Plate IV(3) shows a piece of Scotch tape covering half of a quartz wedge. The slow direction in the quartz wedge is marked with an arrow which is just visible (it is in the direction of the short dimension of the quartz plate). By following any given interference color band from the uncovered area into the tape-covered area, we find that the colors are lower where the tape lies over the quartz wedge. Therefore the "fast direction" in the tape must be parallel to the "slow direction" in

the quartz—i.e., the width direction of the tape is the fast direction and the length is the slow direction. Use the tape to determine the fast and slow directions in cellophane from a cigarette package. First look at the interference color from the tape alone. That shown in Plate IV(3) gives first-order red. Superpose the long direction of the tape over first one vibration direction in the cellophane and then the other (always between crossed polarizers, of course). Are their orientations the same from one package to another? What about plastic film from other sources? Once the fast and slow directions in a piece of cellophane have been identified, it in turn can of course be used as an identifying plate.

7. Many substances that are on hand in most homes form good crystals from water solution, although quite a few of them form very small crystals that are best observed with a hand lens. Examples are aspirin, cream of tartar, epsom salts, table salt, and washing soda.

The paperback book, *Crystals and Crystal Growing,* by Alan Holden and Phylis Singer (Doubleday-Anchor, 1960), gives instructions for growing a number of crystals, some of which are of special interest between crossed polarizers.

12 *Refraction of Light in Crystals*

When a ray of light passes from one substance into a different substance, it usually changes its direction of travel. This fact is essential to the construction of lenses and so is basically responsible for much of our knowledge of outer space and of very small objects and, in fact, for our ability to see at all.

The phenomenon is known as refraction, from the Latin *refractus,* past participle of the verb *refringere,* a compound form coming from the verb *frangere,* meaning *to break.* In some of the older books the word "refringence" is used instead of "refraction," and of course "frangible" is a perfectly good English word meaning "breakable." The light beam does indeed appear to be broken, or at least bent, as in Figs. 12-1 and 12-2. The amount of bending depends upon the relative velocities of the light in the two substances; refraction is therefore closely associated with matters we have been discussing in the preceding chapters. Figs. 12-1 and 12-2 show light passing from air into two dif-

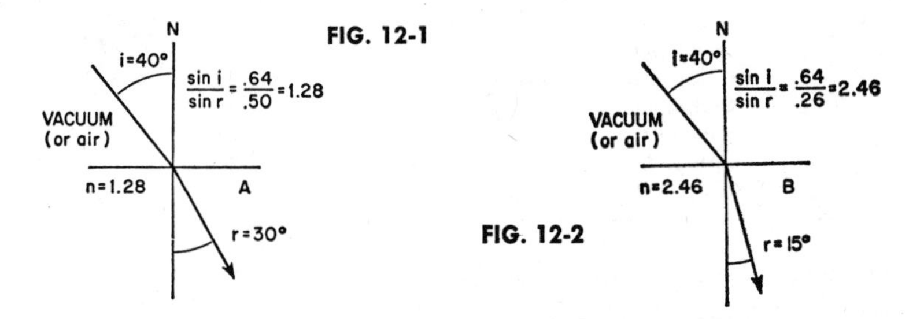

FIG. 12-1

FIG. 12-2

ferent optically isotropic substances. In both cases the light ray meets the surface at an angle of incidence, *i,* of 40°, as measured between the light beam and *N,* the normal to the surface, and some light is reflected, also at an angle of 40° to *N*. In both cases the rest of the light enters the substance, but it is bent more sharply on entering substance *B* than it is on entering substance *A*. Inside of *B*, the ray makes an angle of refraction, *r,* with the normal *N* that measures only 15°, whereas in *A* this angle, *r,* is 30°. The severity of bending of the ray as it enters the substance may be indicated by sin i/sin r, a ratio which is larger for more severe bending and smaller for less severe bending. It is an experimental fact (first discovered by Willebrord Snell, a Dutch astronomer, in 1621) that this ratio remains the same with various angles of incidence and is therefore a constant of the substance. It is called the index of refraction and is denoted by the letter n. For *A* compared to vacuum (Fig. 12-1), $n = 1.28$. For *B* compared to vacuum (Fig. 12-2), $n = 2.46$. As defined, n refers to the ratio sin i/sin r when a ray of light passes from vacuum into the substance in question, but substituting air for vacuum would only require that n be divided by 1.0002, so for our purposes we can substitute air for vacuum with no change of the value of the index. The indices of refraction of a few optically isotropic substances are given in Table 12-1.

Most substances have indices of refraction in the range between 1.3 and 2.5. This may not seem like much of a spread

TABLE 12-1 *Indices of Refraction of Some Optically Isotropic Substances*

Substance	Index of Refraction
Water	1.33
Isopropyl alcohol (the chief constituent of rubbing alcohol)	1.38
Sodium chloride	1.54
Red garnet	1.86
Sphalerite (zinc sulfide)	2.36
Diamond	2.42

when you consider that Figs. 12-1 and 12-2 represent extreme cases, but small differences in index of refraction give readily detectable effects. If rubbing alcohol is poured into water, its course can be followed as it mixes with the water, since the light is bent at the boundary between the two fluids because of their small difference in index of refraction. (See Table 12-1.)* In fact, even the small difference in index of refraction due to small differences in temperature of neighboring air masses above a heater, such as a toaster, is readily detectable.

The "brilliance" of a cut diamond is made possible by its high index of refraction. To understand why, we need to know one more fact about refraction: the ray would travel the same path if it were going in the opposite direction. With the light traveling *from* the high-index substance *into* air, consider what happens to the angle labeled i in Fig. 12-2† as the angle labeled r gets larger. As before, although some light is reflected (less than 5 per cent for r small), much crosses the surface and is refracted.

* To determine the effective refractive index for a substance relative to some other substance that is not air or vacuum, we must divide the index by the index of the other substance. What would the angle of refraction be for a ray going from water into isopropyl alcohol with an angle of incidence of 40°?

$$\frac{\sin 40^\circ}{\sin r} = \frac{1.38}{1.33} = 1.03$$

$$\sin r = \frac{.643}{1.03} = .624 \qquad r = 38.6^\circ$$

† With the light traveling the other way, we should properly reverse the letters for the angles, but it is easier to keep them as they are in the figure.

For some angle of r, i will equal 90°. As i approaches 90°, the intensity of the reflected ray suddenly increases at the expense of the refracted ray. For any angle of r greater than 90° the light will be *totally reflected* back into the substance. The angle r for which the angle i is 90° is called the *critical angle.* By careful cutting of the facets on the bottom side of the diamond, light entering from the top can be made to reflect from the inside of one surface to another and out the top of the diamond again. What is the minimum angle between the surface normal and the ray for total reflection at the inside surface of the diamond? (The answer is given at the end of this chapter.)

The ratio between the velocities of light in two different substances, A and B, is inversely proportional to the ratio between the indices of refraction of the light in these two substances.

$$\frac{V_A}{V_B} = \frac{n_B}{n_A}$$

Fast propagation goes with low index, and slow propagation goes with high index. A mnemonic (memory-aiding) device for recalling which way a ray of light is bent when it goes from a low-index medium into a high-index medium is the pair-of-wheels model in Fig. 12-3. The wheels, joined by an axle, are rolling

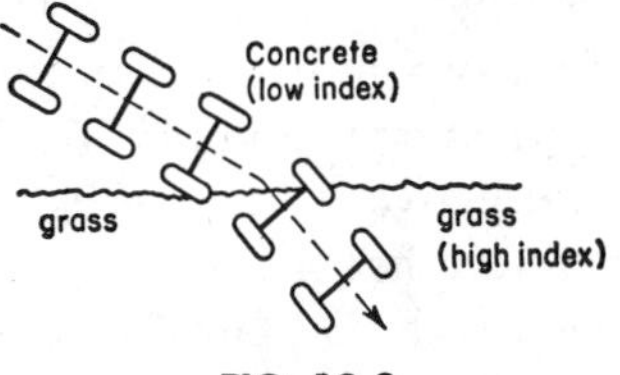

FIG. 12-3

down a gentle slope across a concrete pavement. They meet the boundary between the pavement and some grass at an angle and turn because the grass slows down one of the wheels before the other. It is easy to see that the more the first wheel is retarded by the grass (lower velocity), the greater will be the turning of the direction of travel (higher index of refraction). The more algebraically-minded readers will prefer the equation.

With this relationship between velocity and refractive index to guide us, let us consider the indices of refraction in uniaxial crystals. Clearly there must be more than one index of refraction, since there is more than one velocity of light. Fig. 12-4 is a

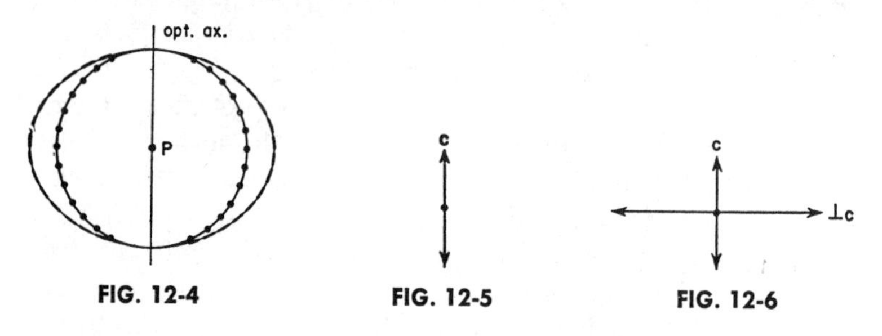

FIG. 12-4 FIG. 12-5 FIG. 12-6

cross section through the familiar ray-velocity surface of a negative uniaxial crystal with the vibration directions shown diagrammatically. Those of the ordinary ray are shown as dots indicating vibration normal to the paper, in this cross section, and the short dashes tangent to the ellipse show the vibration directions of the extraordinary ray, always in the plane defined by the optic axis and the direction of propagation of the ray, and therefore always normal to the vibration of the ordinary ray.

Light originating at point *P* and traveling normal to the optic axis, in the plane of the paper, will, as we know, have two velocities and therefore two indices of refraction, one for light vibrating normal to *c* (the O ray) and the other for light vibrating parallel to *c*. For this negative crystal the light vibrating parallel to *c* is faster and therefore has the lower index.

If we want to make a vector surface for the variation of index of refraction in a uniaxial crystal, it looks as though we should make it as a function of vibration direction. For the case just considered we have a short vector for the lower index of refraction when the vibration direction is parallel to *c* (Fig. 12-5) and a large vector when the vibration direction is normal to *c*, normal to the paper in Fig. 12.4. Since the figure in 12-4 is a cross section of a solid figure in which all sections through the axis of revolution, *c*, are the same, we will do no violence to the truth if we draw the second vector lying in the paper, instead of normal to it, so long as it is normal to *c* (Fig. 12-6). By this reasoning the section of our vector surface normal to *c* must be circular. This circular section of the ellipsoid indicates that for any light traveling along *c*, vibrating in *any* direction normal to its propagation direction, there will be only one index of refraction. This is no surprise, since light traveling along the optic axis has a single velocity. Since the ordinary ray always vibrates normal to the optic axis, this circular section tells all there is to know about the index of refraction of the ordinary ray.

The index of refraction of the extraordinary ray varies from its minimum value for vibration parallel to the optic axis (when the E ray has maximum velocity, as shown in Fig. 12-4) to its maximum value for vibration normal to the optic axis (when the E ray coincides with the O ray). Since the velocity of the E ray varies elliptically with angle to the optic axis, so also, inversely, does its index of refraction. We thus have all the information about the index of refraction of the uniaxial crystal as a function of angle to the optic axis of the vibration direction completely expressed in a single-surfaced ellipsoid of revolution (Fig. 7-6).

We need only two dimensions of such a figure to tell us all about it, once we know that it is an ellipsoid of revolution (as in the thermal expansion case): the equatorial dimension, which is the index of refraction of the ordinary ray, n_{O}; and the axial dimension, which is the extreme index of refraction of the extraordinary ray, n_{E}, the index for the case when the E ray is vibrating parallel to *c* (traveling normal to *c*). These are the two dimensions shown in Fig. 12-6. Index n_{E} is of course smaller than n_{O} for the negative crystal just used as an example and larger than n_{O} for a positive crystal.

If $n_{\mathrm{E}} - n_{\mathrm{O}}$ is $(-)$, the crystal is $(-)$.
If $n_{\mathrm{E}} - n_{\mathrm{O}}$ is $(+)$, the crystal is $(+)$.

This property of having two indices of refraction (which used to be called "refringence," you will recall) is known as *birefringence,* the prefix *bi* needing no explanation to a biped whose grandmother perhaps wears bifocals. The word is also used in a quantitative sense to refer to the numerical difference between the two indices of refraction, as in Table 12-2, where the indices of refraction and birefringence of a few familiar crystals are given. Instead of n_E and n_O, the Greek letters epsilon (ϵ) and omega (ω) are sometimes used. Some authors use small *e* and *o*, and some, but not many, use large N.

TABLE 12-2 *Indices of Refraction of Some Uniaxial Crystals*

Substance	System	n_E	n_O	Birefringence, $n_E - n_O$
Beryl (Aquamarine)	Hexagonal	1.564–1.593	1.568–1.602	(−) very low
Quartz	Trigonal	1.553	1.544	(+) 0.009
Calcite	Trigonal	1.486	1.658	(−) 0.172
Rutile	Tetragonal	2.903	2.616	(+) 0.287

Because of the close relationship between velocity and refractive index, a crystal in which the extreme velocity of the E ray differs only very slightly from that of the O ray has low birefringence, whereas one in which the velocity differences are greater has high birefringence. Suppose we have two specimens of such crystals, both of the same thickness, with large surfaces parallel to the optic axis, lying on a piece of glass between crossed polarizers. The one with the low birefringence will show lower interference colors than the one with higher birefringence. In fact, if the thickness is known, the birefringence can be determined from the interference color, as shown in Fig. 12-7. Thus the difference between the indices of refraction may be measured without a knowledge of either index. Clearly, if any two of these three parameters is known, the third may be determined from the chart.

Suppose we have two other slices from the same two crystals, which, unlike the first two, are cut normal to the optic axis. Between crossed polarizers in parallel light both will show zero birefringence. In convergent light the interference figure of the low-birefringence crystal will show fewer colored rings that that of the high-birefringence crystal, for the same limit of included angle (Fig. 9-2). In Plate V the slice of quartz used for (1a) was 0.35 mm thick, whereas the slice of calcite used for (1b) was

0.13 mm thick, about one third as thick as the quartz. Because of its very high birefringence it shows more rings than the quartz does even though it is a much thinner crystal slice. The fact that even with a small limiting angle (as in viewing without converging lenses) a thin slice of calcite, viewed along the optic axis between crossed polarizers, will show several rings made it suitable for use in the optical ring sight.

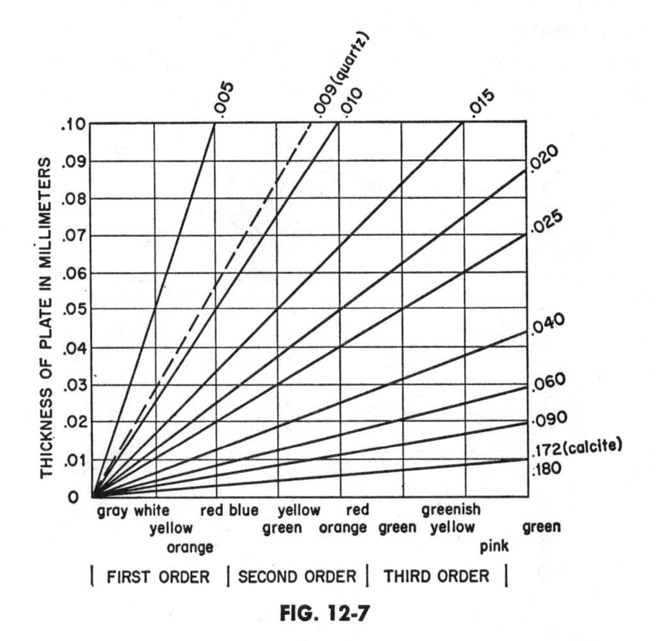

FIG. 12-7

The surface that shows the value of the index of refraction for different vibration directions in a crystal is called *the indicatrix* (plural: indicatrices), or sometimes *the index ellipsoid.* As in the case of the expansion surface for linear thermal expansion, it is a sphere for cubic crystals; an ellipsoid of revolution for tetragonal, hexagonal, and trigonal crystals; and a triaxial ellipsoid for orthorhombic, monoclinic, and triclinic crystals.

Although it would be difficult to measure the velocity of light in a crystal, it is relatively easy to measure the index of refraction. This may be done in several ways, only one of which will be given here. If a small sample of a cubic crystal is immersed in a drop of liquid on a microscope slide, preferably covered with a cover glass, its index of refraction can be compared with that of the liquid. If the two match exactly, it will be impossible to see the crystal fragment unless it is colored or has some flaw or inclusion. If the index of one is very much higher than that of the other, the crystal will be marked by a very distinct, dark boundary line. If the indices are close to each other, a bright line, known as the Becke line, will appear all around the boundary of the crystal. The origin of this line is shown diagrammatically in Fig. 12-8.

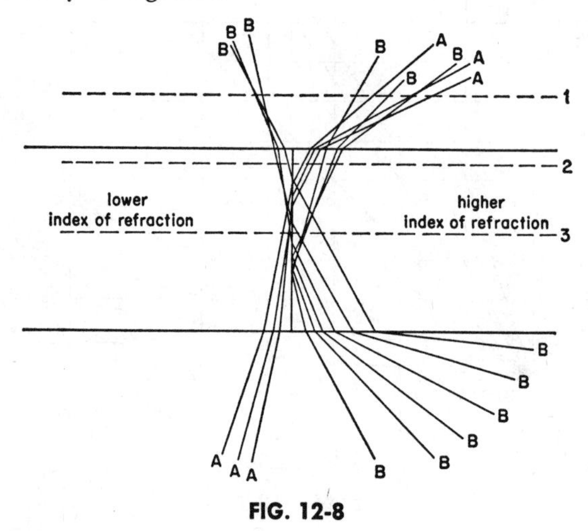

FIG. 12-8

Since the light rays coming up through the crystal are not perfectly parallel, they will meet the crystal-to-liquid boundary at various angles, as shown in Fig. 12-8. Suppose the liquid has the lower index of refraction and the crystal, the higher index. Then most of the rays originating to the left of the boundary in Fig. 12-8 (*A* rays) will be bent into the crystal when they meet the boundary. Some light will be reflected, but much less than on the other side, where some of the rays meeting the boundary from the higher-index side will be totally reflected. The combined effect of refraction of the *A* rays and reflection of the *B* rays will produce a band of greater concentration of light over the crystal (high index) which slopes downward toward the boundary between the crystal and the liquid (lower index). This band can be observed at various levels (1, 2 and 3, in Fig. 12-8) by focusing the microscope at various levels. As the focus is lowered, the bright band moves toward the substance with lower index; as the focus is raised, the bright band moves toward the substance with higher index.

Liquids of known refractive index are available commercially for comparison with crystals in this way. Repeated tests are made (using a clean bit of crystal and fresh slide each time) until the index of the crystal has been found to lie between the indices of two neighboring liquids of the available series of immersion liquids. If a more accurate determination is required, liquids may be especially mixed to achieve the best possible match and the index of refraction of the liquid mixture may then be measured in a refractometer.

In cubic crystals, where the indicatrix is a sphere, one such measurement on a crystal fragment of any orientation in polarized or unpolarized light gives us the index of refraction of the crystal. For very precise measurement, monochromatic light would be used and the temperature of the substances carefully controlled, since the index of refraction can vary both with temperature and with wavelength (see Chapter 14).

In optically uniaxial crystals the index of the O ray and the extreme index of the E ray (smallest if the crystal is negative, largest if it is positive) must be measured separately. A piece of the crystal which can be placed on the stage of the polarizing microscope oriented so that the observer is looking normal to the optic axis should be used (like the section of the tetragonal crystal in Fig. 8-9). Such a piece will show the maximum birefringence, for a given thickness, for light traveling normal to the optic axis, since in this direction the two rays have maximum velocity difference. In order to measure the index of refraction of the O ray, n_O, without any contribution from the E ray, we need only place the vibration direction of the O ray parallel to that of the polarizer, as in Fig. 8-12. In this position there is no component of vibration in the E vibration direction.

To ensure that the O vibration direction is properly aligned, the analyzer is inserted and the crystal rotated (by rotating the stage) back and forth through a decreasing angular range until it is precisely at the extinction (darkest) position. Then the analyzer is removed, and the index of refraction for the O ray is compared with that of the liquid in which it is immersed. The process is repeated for the E ray, using fresh crystals and different liquids. Which is the O and which the E vibration direction must be determined from the shape of the crystal (E parallel to *c* in this section) or from previous determination of the orientation of the optic axis. If the optic sign of the crystal is known, then the E ray may be distinguished from the O ray by the use of accessory plates which will show which vibration direction is that of the faster and which that of the slower ray. (See Chapter 11.)

For crystals in which the indicatrix is a triaxial ellipsoid, two measurements will not suffice. As in the case of the vector surface for linear thermal expansion, the orientation of the index ellipsoid is subject to the following restrictions:

Orthorhombic system: Three mutually perpendicular ellipsoid axes parallel to three crystallographic axes, with no restriction as to which shall be parallel to which.

Monoclinic system: Any one of the three ellipsoid axes parallel to the 2-fold axis of symmetry (parallel to *b*) or normal to the symmetry plane; no restriction on the orientation of the remaining two axes in the *ac* plane.

Triclinic system: No restriction.

One would think that in the triclinic and monoclinic systems several different slices of known orientation would have to be measured in order to determine how the indicatrix was oriented in the crystal. However, we can use the interference figure to tell us the orientation of the indicatrix. The interference figure for these crystals is not uniaxial, but biaxial. Biaxial crystals are discussed in the next chapter.

* * *

Answer to question about the critical angle in diamond: (What is the minimum angle between the surface normal and the ray for total reflection at the inside surface of a diamond?)

Referring to Figs. 12-1 and 12-2 and using 2.42 for the index of refraction of diamond, we need only set i equal to 90° and solve for r.

$$\frac{\sin i}{\sin r} = 2.42$$

$$\sin 90° = 1$$

$$\sin r = 1/2.42 = 0.413$$

$$r = 24.4°$$

Thus a ray that makes an angle of 65.6° or less with the outer surface of the diamond cannot pass through the surface into the air, but is totally reflected! A properly cut diamond looks opaque when viewed from the bottom.

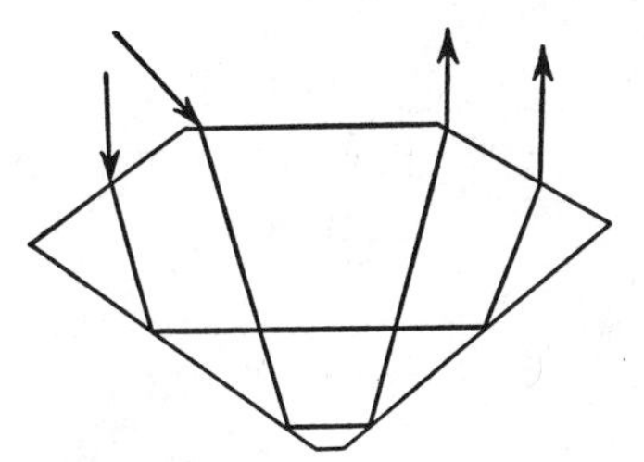

FIG. 12-9 (Adapted from *Gems and Gem Materials*, by E. H. Kraus and C. B. Slawson, 1947. McGraw-Hill Book Co., Inc. Used by permission.)

The accompanying drawing of the cross section of a cut diamond (Fig. 12-9) shows the path of a ray of monochromatic light suffering total reflection at the lower cut surfaces.

13 *Biaxial Crystals*

In all orthorhombic, monoclinic, and triclinic crystals for any one wavelength there are two directions along which the light travels with zero birefringence, as it does along the optic axis of uniaxial crystals. These crystals are therefore said to be *optically biaxial.* The two optic axes define a plane known as the *optic plane,* and the normal to this plane is called the *optic normal* (Fig.

13-1). The angle that the optic axes make with each other may vary from zero to 90°, but it is constant for any particular substance at a given temperature and pressure and for a given wavelength. The direction which bisects the acute angle between the optic axes is called the acute bisectrix (abbreviated Bx_a) and that which bisects the obtuse angle between the optic axes is called the obtuse bisectrix (abbreviated Bx_o).

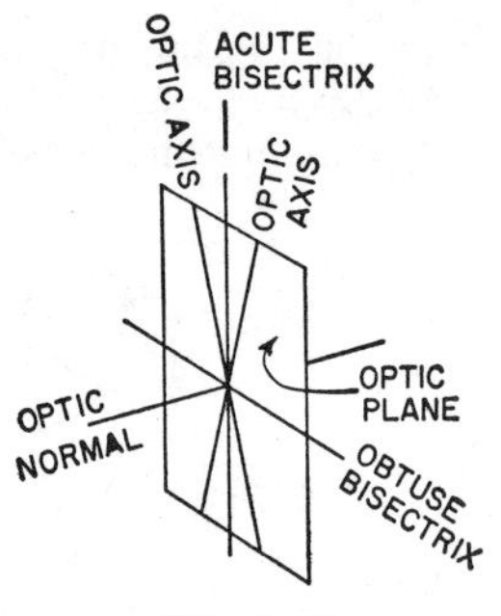

FIG. 13-1

Light traveling in any direction except along an optic axis is broken up into two rays with different velocities and with vibration directions normal to each other. As in the uniaxial case, this retardation, which is zero along an optic axis, increases with angle from the optic axis. In convergent light we therefore see an interference figure with two "eyes" marked by bands of interference colors. In Plate VIII(1) we are looking along the acute bisectrix in the common mica, muscovite. The acute bisectrix is approximately normal to the cleavage in muscovite, and therefore a cleavage flake of the mica will show the acute bisectrix interference figure when viewed in convergent light between crossed polarizers.

Since mica cleaves so easily, pieces of different thicknesses are quickly and easily prepared. Observation of the interference figures from these pieces shows how the bands of equal retardation change appearance according to thickness. The same demonstration could be done with a single piece peeled to various thicknesses, like that in Plate IV(1).

The illustration on the cover of this book belongs at this point in the text. It is the interference figure of muscovite, taken with sodium light.

As in the case of the uniaxial figure, we must know the vibration directions at various points of the figure in order to understand the black areas in Plate VIII(1), which are called *isogyres.* A plot showing approximately the vibration directions in various parts of an acute bisectrix figure like that in Plate VIII(1), is shown in Fig. 13-2, where the two dark spots represent the

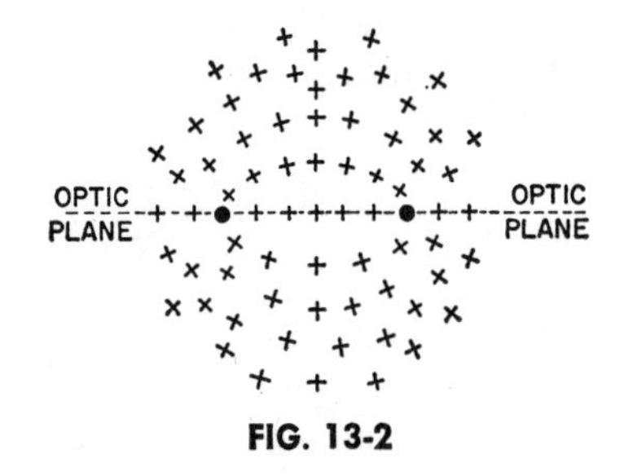

FIG. 13-2

near ends of the two optic axes. As this figure is rotated around the line of sight (normal to the page in Fig. 13-2) between crossed polarizers, which parts will have the vibration directions parallel to those of the Polarizer and Analyzer? Inspection of Fig. 13-2 will show that, unlike the case of the uniaxial crystal viewed along the optic axis (Fig. 9-3), the answer will change as the crystal is rotated. The answers for two different orientations of the crystal (and therefore of the optic plane) are shown in

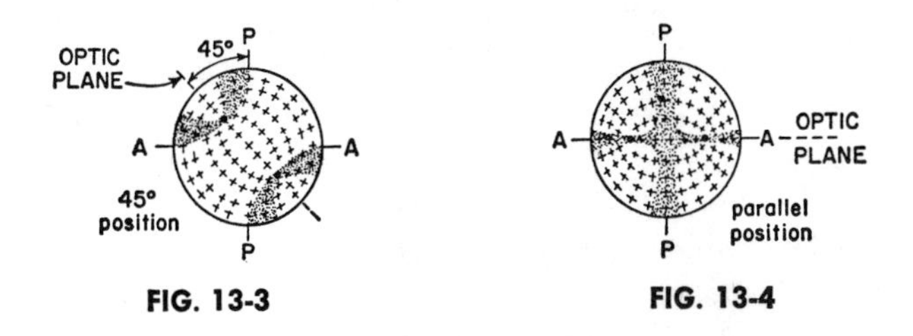

FIG. 13-3

FIG. 13-4

Figs. 13-3 and 13-4. They correspond to the positions of the mica for the interference figures of Plate VIII(1) and VIII(2), respectively. As in all photographs between crossed polarizers in this book, the vibration direction of the Polarizer is up-and-down the page; that of the Analyzer, from left to right. In the first, the optic plane is at 45° to the polarization directions of the Polarizer and Analyzer. In the second, the optic plane is parallel to the vibration direction of the Analyzer. What would be the appearance of the interference figure for a position intermediate between these two? You can determine it with the aid of Fig. 13-2. (The answer is given at the end of this chapter.)

The color you see in the very center of an interference figure in convergent light is the color you would see if you viewed the crystal in parallel light between crossed polarizers, since you would then be looking only along this "central" direction. From this you can see that the "parallel position" of Fig. 13-4 is, in fact, an extinction position of the crystal. In the position shown, if the crystal were viewed in light that was not convergent, with the Polarizer in and the Analyzer removed, all light coming through the crystal would be vibrating parallel to the optic normal, and the index of refraction of the crystal for that vibration direction could be measured in the manner described in Chapter 12.

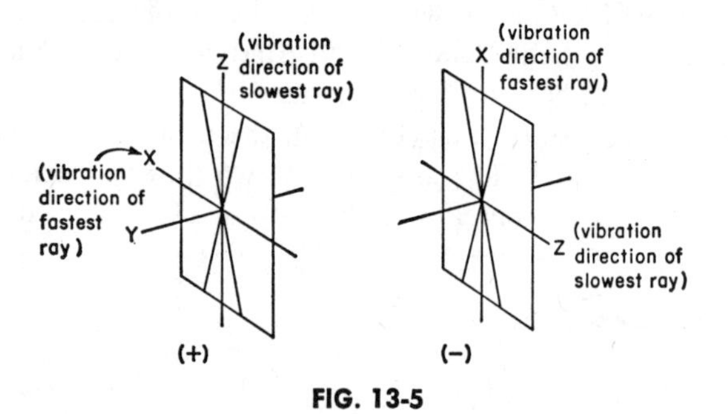

FIG. 13-5

The two bisectrices are the vibration directions of the slowest and fastest rays in the crystal. The vibration direction of the slowest ray is always designated as Z, that of the fastest ray as X. If Z is the acute bisectrix, the crystal is said to be optically positive (+); if X is the acute bisectrix, the crystal is said to be optically negative. (See Fig. 13-5). These two directions are, of course, normal to each other and normal to Y, the optic normal, the vibration direction of a ray of intermediate velocity.

As in the case of the uniaxial crystals, the sign of the crystal can be determined with the aid of an accessory plate. This determination is most conveniently made with the acute bisectrix interference figure.

As we look down the acute bisectrix, the two rays coming toward us through the crystal will be vibrating parallel and normal to Y (see Figs. 13-2 through 13-5)—i.e., normal to the optic plane and parallel to it, respectively. The vibration direction parallel to the optic plane is parallel to the obtuse bisectrix (Fig. 13-1). If this vibration direction tests "fast" compared to Y, the crystal is positive (Fig. 13-5); if it tests "slow," the crystal is negative.

In Plate VIII(3) and VIII(4), the muscovite interference figures are shown, with the addition of the gypsum plate in the accessory slot of the microscope. The plate is in the conventional position with its "slow direction" oriented "NE-SW," or from upper right to lower left. Remembering that the two "eyes" define the orientation of the trace of the optic plane (which passes through them) and comparing the part of the figure between the two isogyres, determine the optic sign of muscovite from Plate VIII(3). (The answer is given at the end of this chapter.)

The X and Z directions, being the vibration directions of the fastest and slowest rays, respectively, are the directions of shortest and longest axes, respectively, of the indicatrix. The third axis of the indicatrix (which is a triaxial ellipsoid for biaxial crystals, as you will recall from Chapter 12) is, of course, normal to the other two and is in the Y direction.

Thus, to determine the indicatrix of biaxial crystals we need only make three measurements, since we can determine with the aid of the interference figure how the X, Y, and Z directions are oriented in the crystal. Then if we can get each of these in turn parallel to the Polarizer, we can measure its index of refraction as described in Chapter 12. Doing so is not always easy. In the mica case, for example, where the acute bisectrix is normal to the cleavage, we must have light traveling along the cleavage flake to measure the index of refraction of the ray vibrating parallel to the acute bisectrix. Standing the flake on edge is not the difficulty. The light will be refracted and reflected at the steep cleavage surfaces which may have caused partings within the flake, and if the flake is too long in the direction of light travel, most of the light will be absorbed. However, few crystals present such grave problems as mica.

The three indices of refraction of a biaxial crystal are known by various names, as given in Table 13-1.

TABLE 13-1 *Indices of Refraction of a Biaxial Crystal*

Index of the fastest ray	= lowest index	$= n_X$	$= n_p$	$= \alpha$
Index of the intermediate ray	= middle index	$= n_Y$	$= n_m$	$= \beta$
Index of the slowest ray	= highest index	$= n_Z$	$= n_g$	$= \gamma$

The subscripts *p*, *m*, and *g* are usually said to stand for "petty," "mean," and "great," but they originally stood for the French "petit," "moyen," and "grand."

Table 13-2 gives the indices of refraction of some biaxial crystals.

TABLE 13-2 *Indices of Refraction of Some Biaxial Crystals*

Substance	Crystal System	α	β	γ
Gypsum ($CaSo_4 \cdot 2H_2O$)	Monoclinic	1.520	1.523	1.530
Microcline (feldspar)	Triclinic	1.522	1.526	1.530
Muscovite	Monoclinic	1.552	1.582	1.588
Sulfur	Orthorhombic	1.950	2.038	2.241

The size of the angle between the two optic axes, which is called *the optic angle,* can be calculated from the refractive indices, as follows:

$$\tan^2 V_Z = \frac{\dfrac{1}{\alpha^2} - \dfrac{1}{\beta^2}}{\dfrac{1}{\beta^2} - \dfrac{1}{\gamma^2}}$$

where V_Z is half the angle which is bisected by Z. if 2V comes out greater than 90°, Z is the obtuse bisectrix, and the crystal is negative. If 2V is less than 90°, Z is the acute bisectrix, and the crystal is positive. Here, as an example, is the calculation of the optic angle for muscovite:

$$\tan^2 V_Z = \frac{\frac{1}{1.552^2} - \frac{1}{1.582^2}}{\frac{1}{1.582^2} - \frac{1}{1.588^2}}$$

$$= \frac{0.4152 - 0.3996}{0.3996 - 0.3966} = \frac{0.0156}{0.0030}$$

$$\tan^2 V_Z = 5.179$$
$$\tan V_Z = 2.2758$$
$$V_Z = 66°18'$$
$$2V_Z = 132°36'$$
$$(-)2V = 47°24', \text{ the optic angle of muscovite.}$$

"(–)2V =" is the conventional way of giving the optic angle of a negative crystal.

The numerical difference, γ minus α, is called the *birefringence* of the crystal. The difference between any other pair of indices is called a *partial birefringence.* The birefringence is measured from a section of a crystal containing X and Z, i.e., from a section parallel to the optic plane. Such a section will show the highest interference colors for a given thickness, since the two rays traveling through it have the maximum velocity difference.

What optical orientations are possible in the various systems? We have already answered this question for the uniaxial crystals where the indicatrix and uniaxial figure have the symmetry ∞/m, $2/m$. (See p. 75)

What is the symmetry of the biaxial figure (Fig. 13-1) and the associated indicatrix which is a triaxial ellipsoid (Fig. 7-7)? Referring to Fig. 13-1, we see that the optic normal is an axis of 2-fold symmetry of the figure; so is the acute bisectrix, and so is the obtuse bisectrix. Normal to each of these three axes is a symmetry plane. So the whole figure has the point-group symmetry *mmm.* Since it has no symmetry axis higher than 2-fold, it is clear that no crystal with more than 2-fold symmetry would be biaxial.

The X, Y, and Z directions (Fig. 13-5) are the three axes of the index ellipsoid. The summary of the possible orientations of these axes given at the end of Chapter 12 therefore applies to these axes of the optic figure.

If a sheet of mica is placed between a pair of crossed polarizers at roughly 45° to the extinction position, the interference figure can be observed with the eye very close to the Analyzer. However, the figure cannot be seen all at once. The polarizer-mica sandwich must be tilted back and forth to let the observer look first along one optic axis and then along the other.

It is in determining the orientation of optically biaxial crystal that a knowledge of optical properties is most powerful. In uniaxial crystals one can determine the orientation of the *c* axis of the crystal by finding the optic axis, but there is no indication of the orientation of the axes that are normal to *c*.

A list of sources of information about optical properties of crystals is given at the end of this book. For a biaxial crystal such references tell not only the orientation of the optic plane relative to the crystallographic axes, but the sign of the crystal, the size of its optic angle, and the orientation of the acute and obtuse bisectrices.

For an example of the use of this sort of information in understanding the interference figure of naphthalene (moth flakes), see Fig. 13-8 and the accompanying text.

Answers to questions in Chapter 13

1. Fig. 13-6 shows the appearance of the interference figure for a position intermediate between that of Fig. 13-3 and that of Fig. 13-4.

Plate VIII(2) and Fig. 13-6 illustrate a fact which every experimenter should keep in mind: that the symmetry of the results can be no greater than the symmetry of the whole experimental system, which includes not only the object investigated but also the means of investigation. The optical properties of the crystal have a symmetry plane both parallel and normal to the optic plane (see Figs. 3-1 and 3-2) and so does Plate VIII(2), where these symmetry elements of the optic figure lie parallel to the same symmetry elements of the crossed polarizers. (The crossed polarizers have point-group symmetry *mm*2.) However, when the crystal is rotated a little away from that coincident position, the only symmetry element which the crystal optics plus the polarizers have left in common is a 2-fold axis along the line of sight. This is the 2-fold axis normal to the paper exhibited by Fig. 13-6.

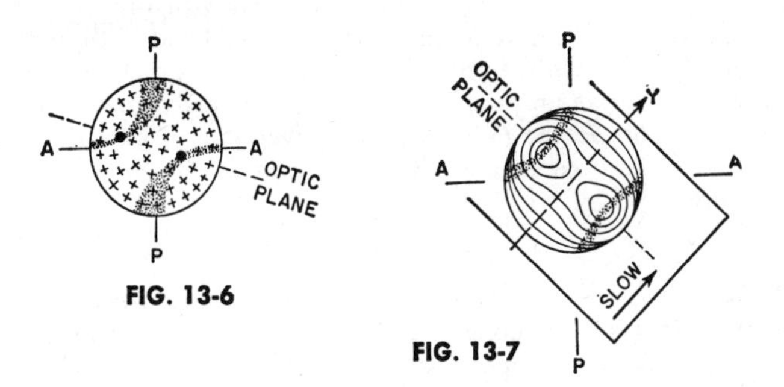

FIG. 13-6

FIG. 13-7

2. Fig. 13-7 shows an analysis of Plate VIII (3). Since the interference colors in the center of the figure are lower colors, because of the use of the accessory plate, "slow" in the plate is over "fast" in the mica crystal. Since we are looking down the acute bisectrix, the optic normal and the obtuse bisectrix (lying in the optic plane) are the vibration directions of the light we are seeing. Since the latter is the slower of the two, it must be Z (compare Fig. 13-5), and the crystal is therefore optically negative.

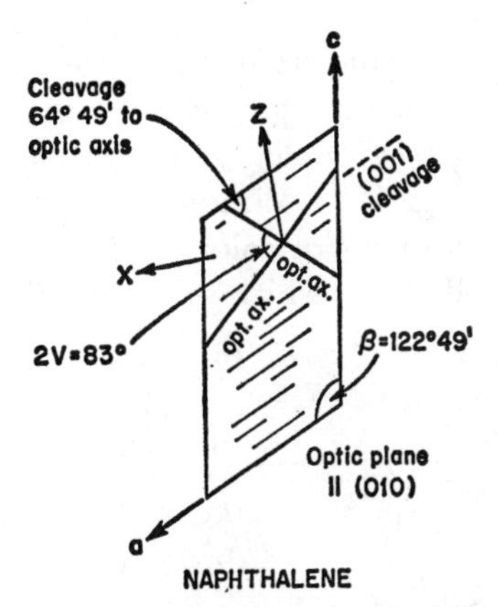

FIG. 13-8 Most flakes of naphthalene (moth flakes) are {001} flakes, their surfaces making an angle of about 65° with an optic axis. The angle between the optic axes is 83°. Between crossed polarizers in convergent light, a flake of naphthalene shows one optic axis close to the edge of the field of view.

When the biaxial figure is in the parallel position with the gypsum plate superposed (Plate VIII(4)), the various quadrants may be used in much the same way as in the case of the uniaxial figure. Note that in this negative biaxial interference figure the black bands indicating exact compensation occur in the upper right and lower left quadrants, whereas those in the positive uniaxial figure of quartz occurred in the upper left and lower right. So the (+) and (—) convention in biaxial crystals is such as to make it consistent with the convention in uniaxial interference figures.

* * *

Problem: Suppose you could look through a biaxial crystal between crossed polarizers right along one of the two optic axes. As you rotated the stage of the microscope, what would you see:

(a) With convergent light?
(b) Without convergent light?

14 *Dispersion*

In most crystals the indices of refraction vary with the wavelength of light. In diamond, for example, the index of refraction is appreciably larger for violet (short wavelength) light than it is for red (long) (Fig. 14-1). White light is therefore broken up into the whole "rainbow" of colors, "violet, indigo, blue, green, yellow, orange and red," which can best be seen when a cut

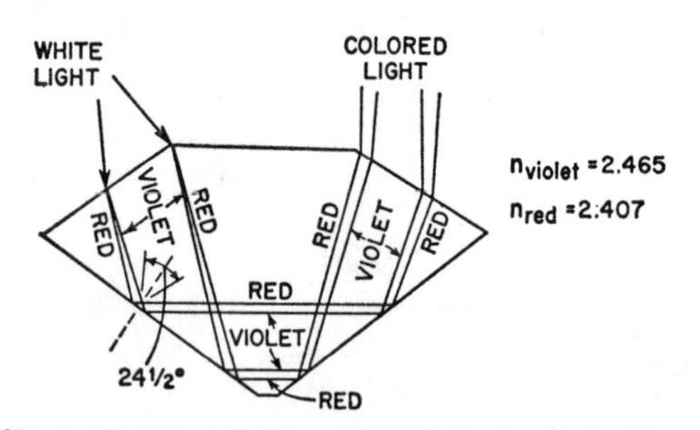

FIG. 14-1 (From *Gems and Gem Materials*, by E. H. Kraus and C. B. Slawson, 1947. McGraw-Hill Book Co., Inc. Used by permission.)

diamond is reflecting sunlight onto a shaded surface. For this reason, when we wanted to show a single ray through the diamond in the chapter on refraction, we had to specify that it was a ray of monochromatic light (Fig. 12-9).

Dispersion is the term applied to the variation of a property with wavelength. Dispersion of the index of refraction in many crystals is too small to measure. In only a very few crystals is it large enough to be noticeable in the course of ordinary optical observation.

The diamond is probably the most striking example of dispersion in optically isotropic crystals. In uniaxial crystals both the O-ray sphere and the E-ray ellipsoid can change with the wavelength of light.

Dispersion in biaxial figures, when large, can produce some spectacular effects in the interference figures. Two examples which are probably the most extreme cases will be mentioned: the monoclinic crystal titanite (also known as sphene, $CaTiSiO_5$) and the orthorhombic crystal brookite (TiO_2). Their indices of refraction for various colors of visible light, where known, are given in Table 14-1.

TABLE 14-1 *Indices of Refraction of Titanite and Brookite for Various Colors of Visible Light**

	α	β	γ	2E†
Titanite				
Green	1.928	1.932	2.064	39°53′
Yellow	1.913	1.921	2.054	45°41′
Red	1.906	1.912	2.041	51°03′
Brookite				
Green	2.627	?	?	33°48′
Yellow	2.583	2.586	2.741	0°
Red	2.541	2.542	2.644	30°47′

* Data from A. N. Winchell, *Elements of Optical Mineralogy.*

† The optic angle *in the crystal* is 2V. The angle measured in air will differ from this because of refraction at the surface and is called 2E. The relation between the two is given by $\sin V = \sin E/\beta$.

The interference figure of titanite is shown in Plate VIII(5) and VIII(6). If titanite were viewed in monochromatic red light, the isogyres would be farther apart (larger optic angle) than they would if the mineral were viewed in monochromatic green light. Wherever the isogyres for a particular color are located, light of that color does not get through. Therefore when the figure is viewed in white light, the place where red doesn't get through looks blue. This is on the outside of the isogyres and farthest from the center, since the optic angle is largest for red. On the inner boundaries of the isogyres we see an orange band where no green gets through, since this is where the isogyres are for green light.

The case of brookite is very special. At room temperature in blue light, brookite is biaxial, with its optic plane parallel to (100). With increasing wavelength, the optic angle decreases until, in yellowish light just a little shorter in wavelength than sodium light, $2V = 0°$. The crystal is uniaxial, with *b* the optic axis. With still longer wavelength light the optic angle opens up again, but in a plane at right angles to its former position; its optic plane is parallel to (001).

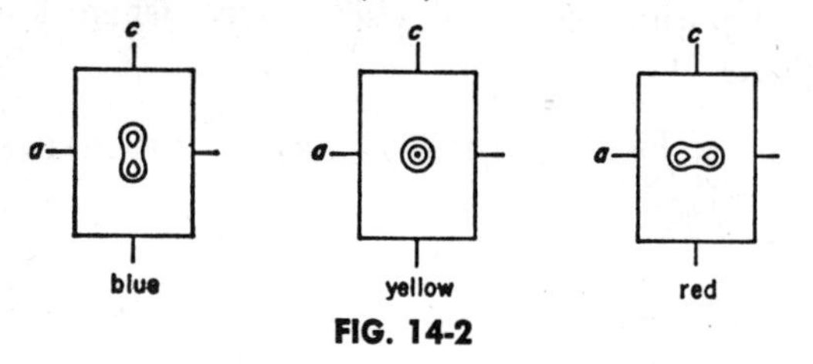

FIG. 14-2

These changes are suggested diagrammatically in Fig. 14-2. If white light is used, all these figures with all the gradations between them combine to give the strange interference figures shown in the photomicrographs, Plate VIII(7) and VIII(8).

The optical properties of most crystals vary with temperature, but this variation has not been widely studied and in most cases is probably small. As in the case of dispersion there are a few remarkable exceptions, but they will not be discussed here.

* * *

Answers to question at the end of Chapter 13.

(a) When you look right along one optic axis of a biaxial crystal in convergent light, you see essentially what you would if you confined your attention to a small circular area around one "eye" of an acute bisectrix figure such as those of Plate V. The appearance of such a figure as the crystal is rotated is shown in Fig. 14-3. Note that the

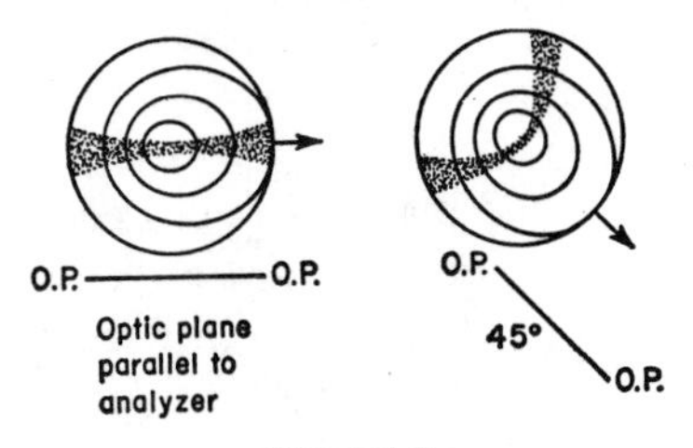

FIG. 14-3

isogyre is straight in the "parallel" positions and curved in the "45° positions." The amount of the curvature depends on the size of the optic angle. If the optic angle is very small, the angle between the "arms" of one isogyre approaches 90° (Fig. 14-4), approaching the ap-

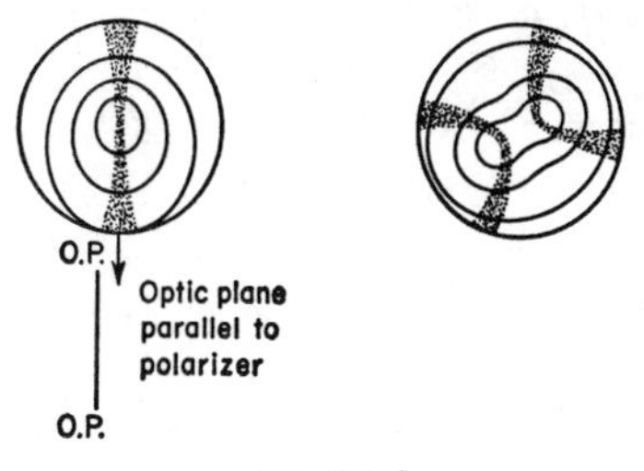

FIG. 14-4

pearance of the uniaxial cross for which the optic angle is zero degrees. In muscovite (Plate V) the optic angle is about 47°. In naphthalene, with an optic angle of 83°, the isogyre is nearly straight, even in the 45° position.

(b) Without convergent light the crystal would remain dark between crossed polarizers as the stage of the microscope is rotated. Note that only in monochromatic light can you look along a single optic axis precisely. Since there is usually a slightly different position for the pair of optic axes for each different wavelength, in white light you see the combined effect, and the crystal looks dark gray (not perfectly dark) as you look along "one optic axis." See Fig. 16-2.

15 *Optical Activity*

Some crystals rotate the plane of polarized light. Light polarized along *P-P* (Fig. 15-1) enters a crystal. Suppose it travels, for example, along the optic axis of a hexagonal crystal. If the crystal rotates the plane of polarized light, then, when the light comes out the other side of the crystal it is vibrating in the direction *C-C,* having been rotated from *P-P* through the angle α. In order to have the Analyzer "crossed" with respect to *C-C,* it too has to be rotated through the angle α. Unless the Analyzer is so rotated, the crystal will not appear dark. If the crystal slice had been thinner, the angle would have been less; if thicker, more. It is as though the vibration direction followed a screw through the crystal, turning a certain number of degrees per millimeter. Consideration of the symmetry of such a screw (Fig. 15-2) shows that this could not occur along the optic axis of a crystal whose optic axis coincided with a symmetry plane, because the screw does not have a symmetry plane parallel to its axis—or normal to it, or anywhere. However, the rotation of the plane of polarized light can occur in crystals with a plane of symmetry if they are biaxial and the rotation along one axis is the mirror image of the rotation along the other (Fig. 15-3). Along one axis the light is acted on in the sense of a right-handed screw, while along the other it is acted on in the sense of a left-handed screw. This whole subject will be discussed further in Chapter 16.

The phenomenon of the rotation of the vibration direction of plane polarized light is known as *rotatory power* or *optical activity*. Substances that do it are said to be *optically active.* Some isotropic substances, such as some cubic crystals and some fluids, do it. In order for a fluid to be optically active its constituent molecules must individually rotate the plane of polarized light since, in a fluid, there is no long-range orderly arrangement. Their randomness of arrangement does not result in cancellation

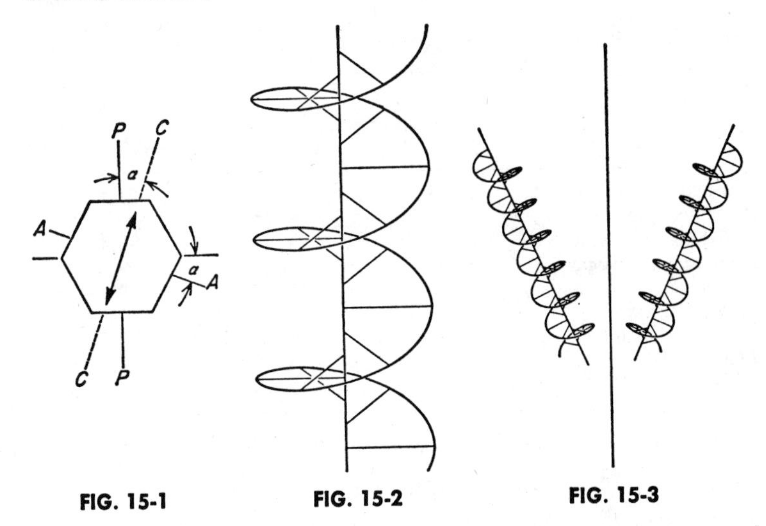

FIG. 15-1 FIG. 15-2 FIG. 15-3

of the effect, since the handedness of a screw is not dependent on orientation. Tables 15-1 and 15-2 give lists of a few optically active substances and their rotatory power in degrees of rotation per path distance.

TABLE 15-1 *Rotation of the Plane of Polarized Light by Some Optically Active Crystals*

Substance	Point Group	Rotation of Sodium Light per Millimeter
Quartz (SiO_2) (along the optic axis)*	32	21°43′
$MgSO_4 \cdot 7H_2O$ (along an optic axis)	222	2°36′
$NaClO_3$	23	3°8′

* Although rotation does occur for light not propagated along an optic axis, the rotation is most easily observed along an optic axis.

Since optical activity is possible in both fluids and cubic crystals, one cannot truthfully say that all optically isotropic substances look black between crossed polarizers. One can truthfully say that all optically isotropic substances affect light the same way regardless of its direction of travel. Regardless of the direction of propagation of light through an optically active isotropic

TABLE 15-2 *Rotation of the Plane of Polarized Light by Some Optically Active Liquids*

Liquid	Concentration (% by wt.)	T (°C)	Rotation of Sodium Light* (Degrees/10 cm path)
Camphor in alcohol	45	20	+21.7
Dextrose *d*-glucose ($C_6H_{12}O_6$) in water	10	20	+5.3
Levulose (fruit sugar) in water	10	25	−9.0
Ethyl *d*-tartrate (Melting point 17°C)	—	20	+7.5
d-Citronellal	—	15	+13.1
l-Citronellal	—	15	−13.1

* + means rotation to the right (clockwise as seen by the observer).

substance, the analyzer would have to be rotated a given amount for a particular thickness to cause extinction of the light.

Like noticeable dispersion of the index of refraction, noticeable optical activity is rare. Probably the most important example of an optically active crystal is quartz, both because it is a widely occurring mineral and because it has such high optical activity. The section of quartz used for the uniaxial figure in Plate V(1) was very thin, only 0.35 mm. thick, as given in Chapter 12, so that the amount of rotation was barely enough to be detectable. If it had been thicker, the cross would not have been as dark in the center.

In most optically active substances, both left-rotating (counterclockwise from the observer's point of view) and right-rotating types occur and both are always possible since the interatomic bonding relationships are in no way different in the two. In quartz crystals, for example, the silica groups are joined to each other in left-handed screw arrangements in some crystals and right-handed screw arrangements in others, the structures being just mirror images of each other. The two can occur together in different regions of a single crystal, and this special kind of twinning is then readily detected between crossed polarizers by rotation of the Analyzer. Some microscopes are equipped with a rotatable Analyzer as described in Chapter 10.

The heading of the third column in Table 15-1 states that

the values given are for rotation of sodium light, suggesting that there exists dispersion of the rotation of the plane of polarized light, which is, in fact, so. In sodium chlorate this dispersion is particularly marked, so that rotation of the analyzer extinguishes one color after another according to its degree of rotation. This succession of colors with rotation of the analyzer is shown in a color plate in the book "Crystals and Crystal Growing" by Holden and Singer (Doubleday Anchor, 1960). It is also so that rotatory power varies with temperature.

16 *Summary of the Relation Between Optical Properties and Symmetry*

We have seen that the symmetry of biaxial optical properties is *mmm*, that of uniaxial properties ∞/m, $2/m$ except in optically active crystals, and that of isotropic properties spherical (with an infinite number of infinite axes and an infinite number of planes)* except in optically active crystals. The optically active crystals will be considered later in this chapter. Consideration of the symmetry of various systems has shown us which sort of optical properties must belong to each and how the optic axes and planes must be oriented. This information is summarized in Table 16-1.

We have also seen that under special conditions a biaxial crystal may become uniaxial. Thus the orthorhombic crystal brookite (Fig. 14-2 and Plate VIII(7 and 8)) is uniaxial for yel-

* Of course the symmetry ∞/m, $2/m$, the symmetry of an ellipsoid of revolution, has an infinite number of planes all through the one infinite axis and 2-fold axes normal to it. The sphere has an infinite number of planes through each of its infinite axes and one normal to it, and it has an infinite number of such axes.

TABLE 16-1 *Optical Orientation in the Various Crystallographic Systems*

Systems	Anisotropic		Isotropic
	Biaxial	Uniaxial	
Triclinic	No orientation restriction		
Monoclinic	Either X, Y or Z parallel to *b*		
Orthorhombic	X, Y, and Z parallel to the crystallographic axes		
Tetragonal Trigonal Hexagonal		Optic axis parallel to *c*	
Cubic			Not orientable

Recall that X, Y and Z are the vibration directions of the fast, medium, and slow rays, respectively; that X and Z are the bisectrices of the angles between the axes and Y the normal to the optic plane.

low light, but this does not mean that it has become a tetragonal crystal. The structure of the crystal has orthorhombic symmetry, as its optical properties for all other colors of light show very well. *The symmetry of a crystal is the symmetry of its least symmetrical property.* If any property of a crystal lacks a 4-fold symmetry axis, then this property reveals to us the fact that the arrangement of atoms in the crystal lacks a 4-fold symmetry axis.

Notice that the optic plane of brookite did not change from (100) to (001) in Fig. 14-2 simply by rotating around the acute bisectrix (normal to the paper). In a crystal with an orthorhombic structure, the optic plane would have to be parallel to the crystallographic axes under all conditions. In monoclinic crystals, the optic directions can be tilted at different angles around the *b* axis for different colors of light, as shown diagrammatically in Figs. 16-1 and 16-2, and this tilting can result in very strange interference figures.

In Chapter 15 the rotation of the plane of polarized light was shown as a phenomenon having the symmetry of a screw. What is the symmetry of a screw? It lacks a center of symmetry. It has no plane of symmetry either parallel to its axis or normal to it. Therefore the rotation of polarized light cannot occur for a light path along any direction parallel or normal to a mirror plane.

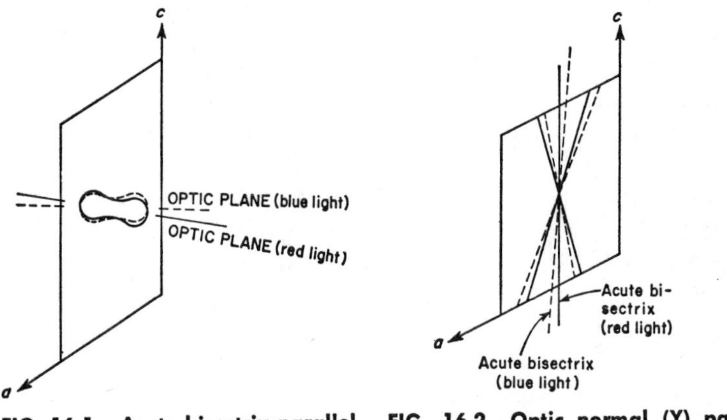

FIG. 16-1 Acute bisectrix parallel to *b*.

FIG. 16-2 Optic normal (Y) parallel to *b*.

Referring to Table 5-2, we see that we have to consider the various point groups of one system separately when we are considering the phenomenon of optical activity. We cannot make a categorical statement about a whole crystal system in the case of optically active crystals as we could for the optical properties discussed above.

Table 16-2 shows the point groups in which rotatory polarization can occur. Although a mirror plane cannot exist parallel or normal to the axis of a screw, a mirror plane at some other

TABLE 16-2 *Crystal Classes in which Optical Activity May Occur*

Triclinic	Monoclinic	Orthorhombic	Tetragonal	Trigonal	Hexagonal	Cubic
1	2	*mm*2	4	32	6	23
	m	222	$\bar{4}$	3	622	432
			422			
			$\bar{4}2m$			

angle to the screw axis simply generates a second screw of opposite handedness to the first one, as shown in Fig. 15-3. In this way, optical activity is possible in crystals with the point groups *m* and *mm*2. In *m* the optic normal must lie in the mirror. In *mm*2 the optic normal must lie along the 2-fold axis.

In biaxial crystals whose optic axes are not related by any symmetry operation of the point group of the crystal, the optical activity along one optic axis does not bear any necessary relation to that along the other axis. Sugar (sucrose) is a good example of such a crystal. Its point group is 2. The optic plane is parallel to (010). Its only symmetry element, the 2-fold axis, is thus normal to the optic plane. Its operation on the optic axes simply turns each through an angle of 180° so that it coincides with itself again. Rotation of polarized light traveling along one of these axes is $-22°$ per cm and rotation along the other is $+64°$ per cm.

In $\bar{4}2m$ we have a very special situation. The (optically uniaxial) crystals with this symmetry cannot rotate the plane of polarized light traveling along the optic axis because two mirror planes pass through this axis. (See Table 5-2.) However, as mentioned briefly in Chapter 15, rotation does occur for light not propagated along the optic axis, and there is nothing in the symmetry $\bar{4}2m$ incompatible with its occurrence provided that it is of opposite handedness in neighboring direction-quadrants, since they are mirror-related. The same sort of analysis applies to the point group $\bar{4}$, since an inversion axis turns a left-handed screw into a right-handed one.

Although it might appear that the same sort of analysis applied to 4*mm*, 3*m*, and 6*mm* would indicate similar off-axis rotation to be possible in these classes, this is not so. Because of the symmetry of these classes, the "components" of right rotation in any direction would be exactly cancelled by "components" of left rotation in the same direction.*

It may seem surprising that a cubic crystal can rotate the plane of polarized light, but let us look at the symmetry elements of point group 23. We see first that if we consider a screw

* In summing the vector components, the squares of the components must be used. (See Wooster or Nye, listed among references at the end of the book.)

along any 2-fold axis (recalling that the symmetry of a screw admits any n-fold symmetry axis along its axis) the point group generates two more along each of the other 2-fold axes. Similarly, if we place a screw along any 3-fold axes, three others spring into existence along the other $\langle 111 \rangle$ directions. If a screw is placed in any general orientation, $[uvw]$, the operations of the 2- and 3-fold axes on it generate 11 other screws, there being 12 such $\langle uvw \rangle$ directions in all, as shown in the diagram of normals of $\{hkl\}$ planes in Table 5-2.

Thus screws in all directions are entirely compatible with the point group 23, and a similar analysis shows them to be compatible with 432. Since they are all related to each other by axes rather than by mirror planes, they are all of the same handedness. Clearly, rotation around a 2-fold axis, regardless of its orientation, cannot change the handedness of a screw. Otherwise we would not be able to put a screw into the ceiling with the same motion we use to put it into the floor!

Notice how very different mirror symmetry and rotational symmetry are. With a boxful of right-handed screws we can arrange them so as to demonstrate any n-fold rotational axis of symmetry (Figs. 16-3 and 16-4). If we want to demonstrate a

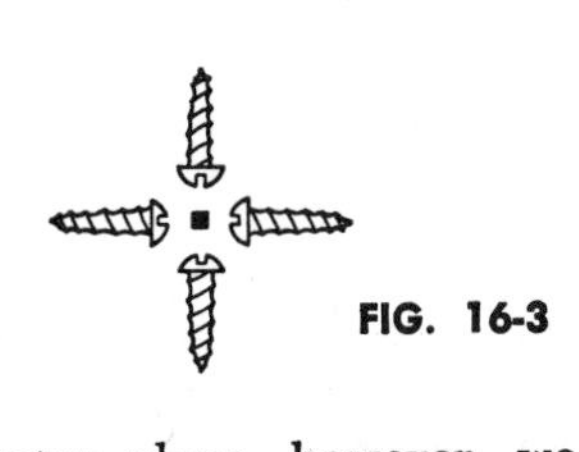

FIG. 16-3

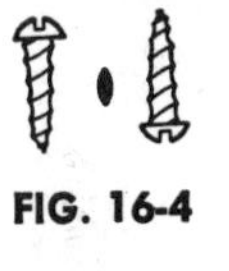

FIG. 16-4

symmetry plane, however, we have to hold a screw up to a mirror and use the optical image; or we can try to buy an exactly similar, but left-handed, screw, which would probably be difficult to find.

This big difference in kind between the symmetry operations that don't change the handedness of the object operated on and those that do has led some authors to speak of them as *symmetry operations of the first kind* and *symmetry operations of the second kind,* respectively.

To which classification does inversion belong: the first

kind, which does not change the handedness, or the second kind, which does? If you hold up your two hands with palms facing each other but with fingers pointing in opposite directions and imagine a point in space, centered between them, you can quickly answer this question. A line from the tip of your left thumb through the center point meets your right thumb-tip at the same distance from the center point on the opposite side. You can repeat the test with various other points on your hands. Inversion is an operation of the second kind: with it you can generate your right hand from your left hand. Rotary inversions are also operations of the second kind.

A screw has no center of symmetry. Could two screws be related by a center of symmetry as they can by a mirror plane (Fig. 15-3)? Careful three-dimensional thinking will show that a right-handed screw can be related to a left-handed screw by a center of symmetry if their axes are parallel. (Note how this is *unlike* Fig. 16-4.) But if light traveling in a given direction is both right-rotated and left-rotated by the same amount, then the net rotation is zero. Crystals with a center of symmetry cannot rotate the plane of polarized light.

17 *Absorption Spectra*

Ordinary white light coming to your eye through a colored crystal, such as a ruby, has lost some of its light in the crystal. Specifically, in this case, the ruby has absorbed light of such wavelengths as to leave the remaining light colored red. Why do some crystals selectively absorb certain wavelengths of light?

We have been speaking of light as "vibrating" in a particular direction as it goes through the crystal. In the vibration direction, it is an oscillating electric field. If, in the crystal, there are some bound charges which can oscillate readily at the same number of oscillations per second as some of the passing light, they will be activated by that electric field and will absorb its energy. How rapidly the charge can oscillate will depend on its mass (an electron is a lot lighter than a whole atom) and the forces of attraction and repulsion exerted on it by its neighbors to hold it in place.

Short-wavelength light acts as a higher-frequency oscillator than long-wavelength light, so the physicist has a whole spectrum of frequencies with which to harass the charges in the crystal. In fact, wavelengths longer than the visible red, on into the infrared, and even microwaves (very-short-wavelength radio, a few centimeters long), can be used if the crystal is transparent to them. On the short-wavelength side, beyond the violet, ultraviolet "light" and even x-rays can be used. We thus learn about the electrons in the crystal from a study of the absorption in the ultraviolet range and about the ions (charged atoms) from studies of the absorption in the infrared range.

When a physicist passes such radiation through a crystal and finds that, when it comes out the other side, radiation of certain well defined frequencies has been absorbed, he then has the problem of making an educated guess, based on his knowledge of the crystal, as to which electrons or ions in the crystal were responding to that frequency and soaking up that energy to stimulate their response. Sometimes he can find the answer and sometimes he can't, but there are many excellent physicists working in this field at the present time who are using light and neighboring radiation as a tool for learning more about the nature of crystals in this way.

The intensity of transmitted light as a function of wavelength for any particular substance is called the *absorption spectrum* (plural, *spectra*) for that substance. It could as well be called the transmission spectrum, but the more common term emphasizes that we are especially interested in those wavelengths for which the transmission is low.

Since the response of the charges in the crystal might be easier in one direction than in another because of the anisotropy of bonding forces, some of these investigators are working with polarized light. They find that the absorption varies with the direction of polarization.

In some crystals this difference in absorption for different vibration directions of the light is so marked that it is readily observable when the crystal is viewed with the Polarizer. The orthorhombic mineral hypersthene looks pink for light transmitted with its vibration along [100], but green for light transmitted with its vibration direction along [001]. This phenomenon is called *dichroism* for two different vibration directions, *pleochroism* when each of the three major directions (X, Y, and Z) shows a different color.

The trigonal mineral tourmaline absorbs most wavelengths of light very strongly for vibration directions normal to the optic axis and very weakly for vibration parallel to the optic axis, so that a slice of tourmaline cut parallel to the optic axis acts somewhat like a polarizing film. In the early days of crystallography two such crystals were mounted in the "crossed" posi-

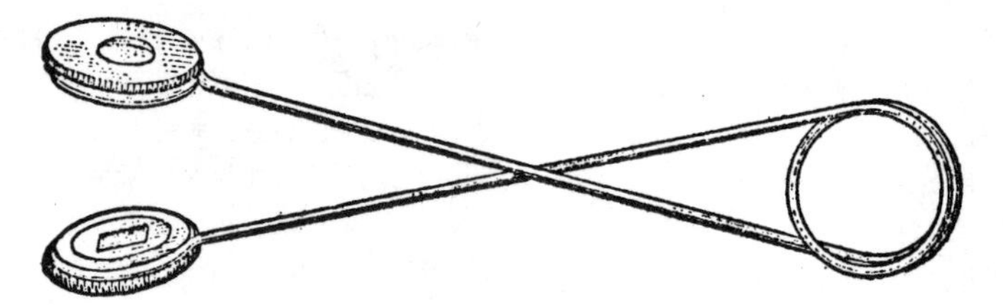

FIG. 17-1 (From *Dana's Textbook of Mineralogy*, edited by W. E. Ford, 4th ed., 1932. By permission of John Wiley and Sons, Inc.)

tion in a pair of "tongs," Fig. 17-1, and one studied crystals in crossed polarized light with "the tourmaline tongs."

The interaction of crystals and light has attracted the attention of students for 300 years. It is a subject being actively investigated today, and it does not appear that we will understand it fully for some time to come.

*Appendix I How They Got Rid of the Black Cross in the Optical Ring Sight**

The optical ring sight, Plate VII, has, in addition to the thin plate of calcite and two sheets of polarizing film, described in Chapter 9, two quarter-wave plates, the whole sandwich being arranged as shown in Fig. I-1.

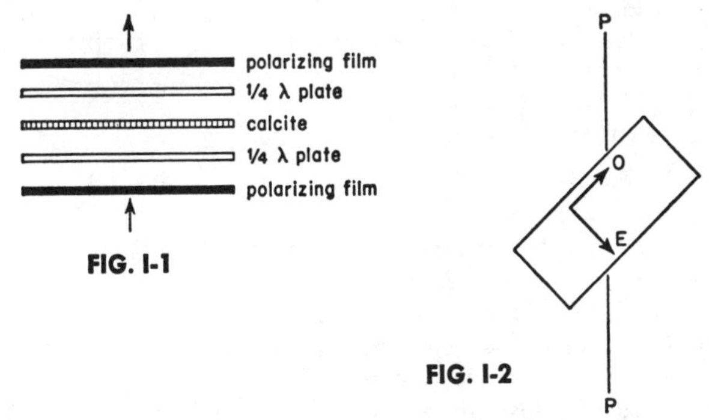

FIG. I-1

FIG. I-2

A quarter-wave plate can be made of any optically anisotropic substance (mica is often used) whose thickness is such that the retardation of one ray with respect to the other is a quarter of a wavelength for the particular color or light in question—for example, sodium light, which is yellow. A quarter-wave plate for yellow light gives a pale gray interference color in white light for either crossed or parallel polarizers.

In the ring sight (Fig. I-1) the quarter-wave plate next to the first polarizer is attached to it with the vibration directions in the plate at 45° to that of the polarizer, the familiar orientation shown in Fig. 8-13. Now what has this done to the light which is about to go through the calcite plate?

* This appendix was critically read by A. Makas and E. Emerson of the Polaroid Corporation, manufacturers of the optical ring sight. Their cooperation in suggesting improvements is gratefully acknowledged.

In the quarter-wave plate the light is broken up into two vibration directions, which we can represent by two vectors of equal length because the plate is at 45° to the vibration direction (*P-P*) of the first polarizer (Fig. I-2). The combined effect of these two motions as the light emerges from the plate, with one of the rays a quarter wavelength behind the other, can be determined in the following way.

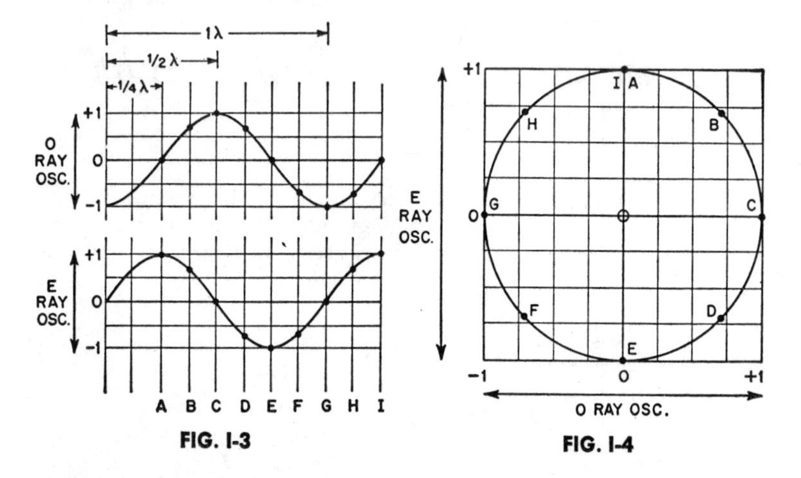

FIG. I-3 FIG. I-4

We can read the contribution of each ray throughout a complete cycle from Fig. I-3. Plotting the oscillation of the E ray along the y axis of Fig. 1-4 and that of the O ray along the x axis, we get the coordinates of a set of points representing the combined motion.

These points lie on a circle, and the light is therefore called *circularly polarized light.* It has lost any record of the polarization direction of the first polarizer.

From Fig. 9-3 we recall that the black cross resulted where vibration directions within the crystal were parallel to either the polarization direction of the Polarizer or that of the Analyzer. We have seen how the polarization direction of the Polarizer got lost before the light entered the calcite crystal. In the calcite crystal the light will be broken up as always into the ordinary and extraordinary rays, and these will have the vibration directions familiar to us. In the case of convergent light, the pattern of vibration directions will be as shown in Fig. 9-3. The retardation of one ray with respect to the other will result in concentric bands of retardation increasing outward from the center as before.

Following the calcite is a second quarter-wave plate with the vibration directions of its slow and fast rays parallel to those of the first. It cannot give circularly polarized light everywhere because, in order to do so, it must receive plane polarized light vibrating at 45° to its vibration directions (Fig. I-2). What it does do is analyze each different vibration direction of Fig. 9-3 into two components parallel to the two (mutually perpendicular) vibration directions of this second quarter-wave plate.

Following this second plate is the second polarizer, or Analyzer, crossed with respect to the first, the Polarizer. Since this is at 45° to the two components it now receives, light gets through this Analyzer.

In the center of the interference figure, where the light travels along or nearly along the optic axis, it is unaffected by the calcite, and the result is just the combined retardation of the two quarter-wave plates. Each one alone gives a gray interference color (see Plate IV(2), a quarter of the way along the First-Order sequence). The two retardations add up to a half-wave retardation, giving white in the center of the rings (see Plate IV (2), half way along the First-Order sequence).

Had the second quarter-wave plate been rotated 90° around the light path relative to the first, giving "slow over fast" and "fast over slow," the retardation in the second would just have cancelled the retardation in the first. The center of the figure would have appeared black—that is, no light would get through—and the device would be useless as a "sight."

It is the two parallel quarter-wave plates at either side of the calcite that cause the disappearance of the black cross.

Since the quarter-wave plates cause $\frac{1}{4}\lambda$ retardation only for one particular wavelength, one wonders how they can work for white light, since for other wavelengths the retardation will not be exactly a quarter wavelength.*

Actually they don't work perfectly throughout the whole range of wavelengths of visible light, so the colored rings are a bit fuzzy. In order to sharpen the rings, the makers have included a filter disc (the dark disc in Plate VII (1)) which can be inserted in the light path so that only that narrow band of wavelengths is observed for which the quarter-wave plate is made. This filter disc greatly sharpens the rings.

* When the retardation is just half a wavelength, the result is plane polarized light—that is, the figure derived in the same way as that in Fig. I-4 would be a straight line with a 45° slope. For any intermediate fraction the figure is an ellipse.

Appendix II Space Groups

When we consider in detail the symmetry of the arrangement of atoms in a crystal, we find that the position of any particular symmetry element in space becomes important. In the calcite-structure pattern on the inside of the front cover, for example, the 3-fold axis normal to the paper cannot be placed at random and still be valid. Only certain points have 3-fold symmetry. Similarly, in the sodium chloride structure in Fig. 4-28, every row of atoms has a 4-fold axis along it, but there are no 4-fold axes between the rows. Such an array of symmetry elements in space which is self-consistent in its symmetry operations is called a *space group.*

In some space groups additional symmetry elements are used to describe the relations among the atom positions. These elements represent the combination of translation with the symmetry operations we already know. Reflection across a plane, combined with a translation of half a unit length along an axis parallel to that plane, is represented by a *glide plane* (Fig. II-1). Translation combined with rotation is repre-

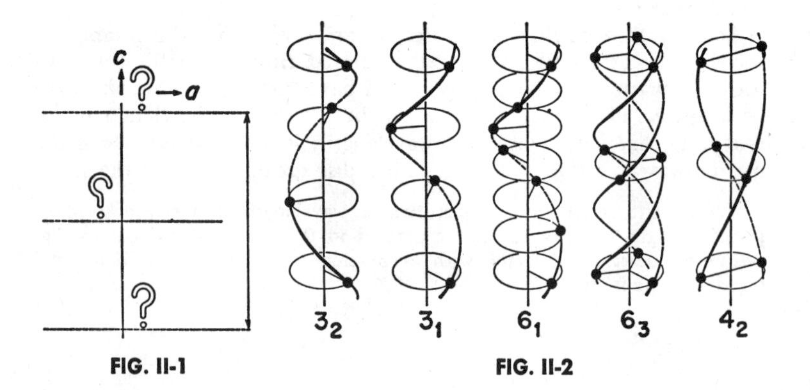

FIG. II-1 **FIG. II-2**

sented by a *screw axis.* Several screw axes are shown in Fig. II-2. The total number of possible self-consistent arrays of symmetry elements in space, including glide planes and screw axes, is 230.

In the symbol of the space group a letter *a, b,* or *c* tells the *direction* of glide of the glide plane, while the position of the letter in the symbol tells which axis the plane is normal to. For example, if an orthorhombic crystal of point group *mmm* had the glide plane shown in Figure II-1, i.e., normal to *b* with glide in the *c* direction—this would be indicated by *mcm*. The letter *n* indicates a diagonal glide, with translation of ½ the cell length in each of two directions; the letter *d* a diamond glide with translation of ¼ the cell length in each of two directions (three in the cubic system). It is also customary to indicate in the space-group symbol whether the space lattice (Fig. 2-25) is primitive, *P* (with lattice points only at the corners); face-centered, *F* (with lattice points in the centers of all faces), *B* (with lattice points in the center of {010} only, etc.); or body-centered, *I* (German, *Innere*). A screw axis is indicated by a numerical subscript, *s*, which indicates that the next atom up along the *n*-fold screw is s/n of the total length of the cell, measured along the screw axis (see Fig. II-2). Rotation is counterclockwise, as seen from the top.

Except in discussing the arrangement of atoms in the structure or the x-ray reflections resulting from this arrangement, we do not need to concern ourselves with space groups and can confine our attention to the point group.

Given the space-group symbol of a crystal, one can determine the point group by converting all the glide planes to regular mirror planes and all the screw axes to regular rotation axes. The space-lattice letter is also not used for the point group. Table II-1 gives a few examples.

TABLE II-1 *Space Group and Point Group of Selected Substances*

Substance	Space Group	Point Group
Aluminum (Al)	*Fm3m*	*m3m*
Diamond (C)	*Fd3m*	*m3m*
Molybdenum (Mo)	*Im3m*	*m3m*
Calcite ($CaCO_3$)	$R\bar{3}c$*	$\bar{3}m$
Cesium chloride (CsCl)	*Pm3m*	*m3m*
Beryl ($Be_3Al_2Si_6O_{18}$)	$P6/mcc$	$6/mmm$
Cobalt (Co)	$P6_3/mmc$	$6/mmm$
Quartz (SiO_2)	$P3_121$ or $P3_221$	32
Barite ($BaSO_4$)	*Pbnm*	*mmm*

* *R* indicates a rhombohedral lattice, see Fig. 2-24(9).

Bibliography

1. SYMMETRY

Harold Hilton, *Mathematical Crystallography,* Clarendon Press, Oxford, England, 1903. Unabridged, corrected edition; Dover, New York, 1963.

D'Arcy W. Thompson, *On Growth and Form,* Cambridge University Press, 1942. Two volumes.

H. Weyl, *Symmetry,* Princeton University Press, 1952.

2. CRYSTALLOGRAPHY AND CRYSTALLOGRAPHERS

C. S. Barrett, *Structure of Metals,* McGraw-Hill, New York, 1952.

F. D. Bloss, *An Introduction to the Methods of Optical Crystallography,* Holt, Rinehart and Winston, New York, 1961.

C. V. Boys, *Soap Bubbles, Thin Colours and the Forces which Mould Them,* Dover, New York, 1959.

C. W. Bunn, *Chemical Crystallography,* Oxford University Press, Second edition, 1963.

E. S. Dana, *A Textbook of Mineralogy,* revised and enlarged by W. E. Ford, Wiley, New York, 1932.

P. P. Ewald, *Fifty Years of X-ray Diffraction,* published for the International Union of Crystallography by N. V. A. Oosthoek's Uitgeversmaatschappij, Utrecht, The Netherlands.

N. H. Hartshorne and A. Stuart, *Crystals and the Polarizing Microscope,* Edward Arnold and Co., London, Third edition, 1960.

N. F. M. Henry and Kathleen Lonsdale, editors, *International Tables for X-ray Crystallography,* published for the International Union of Crystallography by the Kynoch Press, Birmingham, England, 1952. See especially the Historical Introduction by M. von Laue, pp. 1-5, and Point-group Symmetry and the Physical Properties of Crystals, pp. 41-43.

F. A. Jenkins and E. White, *Fundamentals of Optics,* McGraw-Hill, New York, Second edition, 1950, Chapters 1, 26, 28.

A. Johannsen, *Manual of Petrographic Methods,* McGraw-Hill, New York, 1918.

P. F. Kerr, *Optical Mineralogy,* McGraw-Hill, New York, Third edition, 1959. See especially the full-color chart of interference colors.

K. Lonsdale, *Crystals and X-rays,* Bell, London, 1948.

J. F. Nye, *Physical Properties of Crystals,* Clarendon Press, Oxford, 1957.

F. C. Phillips, *An Introduction to Crystallography,* Longmans, Green, New York, 1956.

A. V. Shubnikov, *Principles of Optical Crystallography.* Original Russian text published for the Institute of Crystallography by the Academy of Sciences, U.S.S.R. Press, Moscow, 1958. English translation, Consultants Bureau, New York, 1960.

E. E. Wahlstrom, *Optical Crystallography,* Wiley, New York, Third edition, 1960.

A. G. Ward, *The Nature of Crystals,* Blackie and Son, Glasgow, 1938.

E. A. Wood, *Crystal Orientation Manual,* Columbia University Press, 1963.

W. A. Wooster, *Text-book on Crystal Physics,* Cambridge University Press, 1938.

Index

A CATALOGUE OF SELECTED DOVER BOOKS IN ALL FIELDS OF INTEREST

A CATALOGUE OF SELECTED DOVER BOOKS IN ALL FIELDS OF INTEREST

CREATIVE LITHOGRAPHY AND HOW TO DO IT, Grant Arnold. Lithography as art form: working directly on stone, transfer of drawings, lithotint, mezzotint, color printing; also metal plates. Detailed, thorough. 27 illustrations. 214pp.
21208-4 Pa. $3.00

DESIGN MOTIFS OF ANCIENT MEXICO, Jorge Enciso. Vigorous, powerful ceramic stamp impressions — Maya, Aztec, Toltec, Olmec. Serpents, gods, priests, dancers, etc. 153pp. 6⅛ x 9¼. 20084-1 Pa. $2.50

AMERICAN INDIAN DESIGN AND DECORATION, Leroy Appleton. Full text, plus more than 700 precise drawings of Inca, Maya, Aztec, Pueblo, Plains, NW Coast basketry, sculpture, painting, pottery, sand paintings, metal, etc. 4 plates in color. 279pp. 8⅜ x 11¼. 22704-9 Pa. $4.50

CHINESE LATTICE DESIGNS, Daniel S. Dye. Incredibly beautiful geometric designs: circles, voluted, simple dissections, etc. Inexhaustible source of ideas, motifs. 1239 illustrations. 469pp. 6⅛ x 9¼. 23096-1 Pa. $5.00

JAPANESE DESIGN MOTIFS, Matsuya Co. Mon, or heraldic designs. Over 4000 typical, beautiful designs: birds, animals, flowers, swords, fans, geometric; all beautifully stylized. 213pp. 11⅜ x 8¼. 22874-6 Pa. $5.00

PERSPECTIVE, Jan Vredeman de Vries. 73 perspective plates from 1604 edition; buildings, townscapes, stairways, fantastic scenes. Remarkable for beauty, surrealistic atmosphere; real eye-catchers. Introduction by Adolf Placzek. 74pp. 11⅜ x 8¼. 20186-4 Pa. $2.75

EARLY AMERICAN DESIGN MOTIFS, Suzanne E. Chapman. 497 motifs, designs, from painting on wood, ceramics, appliqué, glassware, samplers, metal work, etc. Florals, landscapes, birds and animals, geometrics, letters, etc. Inexhaustible. Enlarged edition. 138pp. 8⅜ x 11¼. 22985-8 Pa. $3.50
23084-8 Clothbd. $7.95

VICTORIAN STENCILS FOR DESIGN AND DECORATION, edited by E.V. Gillon, Jr. 113 wonderful ornate Victorian pieces from German sources; florals, geometrics; borders, corner pieces; bird motifs, etc. 64pp. 9⅜ x 12¼. 21995-X Pa. $2.75

ART NOUVEAU: AN ANTHOLOGY OF DESIGN AND ILLUSTRATION FROM THE STUDIO, edited by E.V. Gillon, Jr. Graphic arts: book jackets, posters, engravings, illustrations, decorations; Crane, Beardsley, Bradley and many others. Inexhaustible. 92pp. 8⅛ x 11. 22388-4 Pa. $2.50

ORIGINAL ART DECO DESIGNS, William Rowe. First-rate, highly imaginative modern Art Deco frames, borders, compositions, alphabets, florals, insectals, Wurlitzer-types, etc. Much finest modern Art Deco. 80 plates, 8 in color. 8⅜ x 11¼. 22567-4 Pa. $3.50

HANDBOOK OF DESIGNS AND DEVICES, Clarence P. Hornung. Over 1800 basic geometric designs based on circle, triangle, square, scroll, cross, etc. Largest such collection in existence. 261pp. 20125-2 Pa. $2.75

VICTORIAN HOUSES: A TREASURY OF LESSER-KNOWN EXAMPLES, Edmund Gillon and Clay Lancaster. 116 photographs, excellent commentary illustrate distinct characteristics, many borrowings of local Victorian architecture. Octagonal houses, Americanized chalets, grand country estates, small cottages, etc. Rich heritage often overlooked. 116 plates. 11$^{3}/_{8}$ x 10. 22966-1 Pa. $4.00

STICKS AND STONES, Lewis Mumford. Great classic of American cultural history; architecture from medieval-inspired earliest forms to 20th century; evolution of structure and style, influence of environment. 21 illustrations. 113pp. 20202-X Pa. $2.50

ON THE LAWS OF JAPANESE PAINTING, Henry P. Bowie. Best substitute for training with genius Oriental master, based on years of study in Kano school. Philosophy, brushes, inks, style, etc. 66 illustrations. 117pp. 6$^{1}/_{8}$ x 9¼. 20030-2 Pa. $4.50

A HANDBOOK OF ANATOMY FOR ART STUDENTS, Arthur Thomson. Virtually exhaustive. Skeletal structure, muscles, heads, special features. Full text, anatomical figures, undraped photos. Male and female. 337 illustrations. 459pp. 21163-0 Pa. $5.00

AN ATLAS OF ANATOMY FOR ARTISTS, Fritz Schider. Finest text, working book. Full text, plus anatomical illustrations; plates by great artists showing anatomy. 593 illustrations. 192pp. 7$^{7}/_{8}$ x 10¾. 20241-0 Clothbd. $6.95

THE HUMAN FIGURE IN MOTION, Eadweard Muybridge. More than 4500 stopped-action photos, in action series, showing undraped men, women, children jumping, lying down, throwing, sitting, wrestling, carrying, etc. "Unparalleled dictionary for artists," American Artist. Taken by great 19th century photographer. 390pp. 7$^{7}/_{8}$ x 10$^{5}/_{8}$. 20204-6 Clothbd. $12.50

AN ATLAS OF ANIMAL ANATOMY FOR ARTISTS, W. Ellenberger et al. Horses, dogs, cats, lions, cattle, deer, etc. Muscles, skeleton, surface features. The basic work. Enlarged edition. 288 illustrations. 151pp. 9$^{3}/_{8}$ x 12¼. 20082-5 Pa. $4.50

LETTER FORMS: 110 COMPLETE ALPHABETS, Frederick Lambert. 110 sets of capital letters; 16 lower case alphabets; 70 sets of numbers and other symbols. Edited and expanded by Theodore Menten. 110pp. 8$^{1}/_{8}$ x 11. 22872-X Pa. $3.00

THE METHODS OF CONSTRUCTION OF CELTIC ART, George Bain. Simple geometric techniques for making wonderful Celtic interlacements, spirals, Kells-type initials, animals, humans, etc. Unique for artists, craftsmen. Over 500 illustrations. 160pp. 9 x 12. USO 22923-8 Pa. $4.00

SCULPTURE, PRINCIPLES AND PRACTICE, Louis Slobodkin. Step by step approach to clay, plaster, metals, stone; classical and modern. 253 drawings, photos. 255pp. 8$^{1}/_{8}$ x 11. 22960-2 Pa. $5.00

THE ART OF ETCHING, E.S. Lumsden. Clear, detailed instructions for etching, drypoint, softground, aquatint; from 1st sketch to print. Very detailed, thorough. 200 illustrations. 376pp. 20049-3 Pa. $3.75

CONSTRUCTION OF AMERICAN FURNITURE TREASURES, Lester Margon. 344 detail drawings, complete text on constructing exact reproductions of 38 early American masterpieces: Hepplewhite sideboard, Duncan Phyfe drop-leaf table, mantel clock, gate-leg dining table, Pa. German cupboard, more. 38 plates. 54 photographs. 168pp. 8⅜ x 11¼. 23056-2 Pa. $4.00

JEWELRY MAKING AND DESIGN, Augustus F. Rose, Antonio Cirino. Professional secrets revealed in thorough, practical guide: tools, materials, processes; rings, brooches, chains, cast pieces, enamelling, setting stones, etc. Do not confuse with skimpy introductions: beginner can use, professional can learn from it. Over 200 illustrations. 306pp. 21750-7 Pa. $3.00

METALWORK AND ENAMELLING, Herbert Maryon. Generally conceded best all-around book. Countless trade secrets: materials, tools, soldering, filigree, setting, inlay, niello, repoussé, casting, polishing, etc. For beginner or expert. Author was foremost British expert. 330 illustrations. 335pp. 22702-2 Pa. $3.50

WEAVING WITH FOOT-POWER LOOMS, Edward F. Worst. Setting up a loom, beginning to weave, constructing equipment, using dyes, more, plus over 285 drafts of traditional patterns including Colonial and Swedish weaves. More than 200 other figures. For beginning and advanced. 275pp. 8¾ x 6⅜. 23064-3 Pa. $4.50

WEAVING A NAVAJO BLANKET, Gladys A. Reichard. Foremost anthropologist studied under Navajo women, reveals every step in process from wool, dyeing, spinning, setting up loom, designing, weaving. Much history, symbolism. With this book you could make one yourself. 97 illustrations. 222pp. 22992-0 Pa. $3.00

NATURAL DYES AND HOME DYEING, Rita J. Adrosko. Use natural ingredients: bark, flowers, leaves, lichens, insects etc. Over 135 specific recipes from historical sources for cotton, wool, other fabrics. Genuine premodern handicrafts. 12 illustrations. 160pp. 22688-3 Pa. $2.00

THE HAND DECORATION OF FABRICS, Francis J. Kafka. Outstanding, profusely illustrated guide to stenciling, batik, block printing, tie dyeing, freehand painting, silk screen printing, and novelty decoration. 356 illustrations. 198pp. 6 x 9. 21401-X Pa. $3.00

THOMAS NAST: CARTOONS AND ILLUSTRATIONS, with text by Thomas Nast St. Hill. Father of American political cartooning. Cartoons that destroyed Tweed Ring; inflation, free love, church and state; original Republican elephant and Democratic donkey; Santa Claus; more. 117 illustrations. 146pp. 9 x 12.
22983-1 Pa. $4.00
23067-8 Clothbd. $8.50

FREDERIC REMINGTON: 173 DRAWINGS AND ILLUSTRATIONS. Most famous of the Western artists, most responsible for our myths about the American West in its untamed days. Complete reprinting of *Drawings of Frederic Remington* (1897), plus other selections. 4 additional drawings in color on covers. 140pp. 9 x 12. 20714-5 Pa. $3.95

EARLY NEW ENGLAND GRAVESTONE RUBBINGS, Edmund V. Gillon, Jr. 43 photographs, 226 rubbings show heavily symbolic, macabre, sometimes humorous primitive American art. Up to early 19th century. 207pp. 8³/8 x 11¼.
21380-3 Pa. $4.00

L.J.M. DAGUERRE: THE HISTORY OF THE DIORAMA AND THE DAGUERREOTYPE, Helmut and Alison Gernsheim. Definitive account. Early history, life and work of Daguerre; discovery of daguerreotype process; diffusion abroad; other early photography. 124 illustrations. 226pp. 6¹/6 x 9¼. 22290-X Pa. $4.00

PHOTOGRAPHY AND THE AMERICAN SCENE, Robert Taft. The basic book on American photography as art, recording form, 1839-1889. Development, influence on society, great photographers, types (portraits, war, frontier, etc.), whatever else needed. Inexhaustible. Illustrated with 322 early photos, daguerreotypes, tintypes, stereo slides, etc. 546pp. 6¹/8 x 9¼. 21201-7 Pa. $5.95

PHOTOGRAPHIC SKETCHBOOK OF THE CIVIL WAR, Alexander Gardner. Reproduction of 1866 volume with 100 on-the-field photographs: Manassas, Lincoln on battlefield, slave pens, etc. Introduction by E.F. Bleiler. 224pp. 10¾ x 9.
22731-6 Pa. $5.00

THE MOVIES: A PICTURE QUIZ BOOK, Stanley Appelbaum & Hayward Cirker. Match stars with their movies, name actors and actresses, test your movie skill with 241 stills from 236 great movies, 1902-1959. Indexes of performers and films. 128pp. 8³/8 x 9¼. 20222-4 Pa. $2.50

THE TALKIES, Richard Griffith. Anthology of features, articles from Photoplay, 1928-1940, reproduced complete. Stars, famous movies, technical features, fabulous ads, etc.; Garbo, Chaplin, King Kong, Lubitsch, etc. 4 color plates, scores of illustrations. 327pp. 8³/8 x 11¼. 22762-6 Pa. $6.95

THE MOVIE MUSICAL FROM VITAPHONE TO "42ND STREET," edited by Miles Kreuger. Relive the rise of the movie musical as reported in the pages of Photoplay magazine (1926-1933): every movie review, cast list, ad, and record review; every significant feature article, production still, biography, forecast, and gossip story. Profusely illustrated. 367pp. 8³/8 x 11¼. 23154-2 Pa. $7.95

JOHANN SEBASTIAN BACH, Philipp Spitta. Great classic of biography, musical commentary, with hundreds of pieces analyzed. Also good for Bach's contemporaries. 450 musical examples. Total of 1799pp.
EUK 22278-0, 22279-9 Clothbd., Two vol. set $25.00

BEETHOVEN AND HIS NINE SYMPHONIES, Sir George Grove. Thorough history, analysis, commentary on symphonies and some related pieces. For either beginner or advanced student. 436 musical passages. 407pp. 20334-4 Pa. $4.00

MOZART AND HIS PIANO CONCERTOS, Cuthbert Girdlestone. The only full-length study. Detailed analyses of all 21 concertos, sources; 417 musical examples. 509pp. 21271-8 Pa. $6.00

THE FITZWILLIAM VIRGINAL BOOK, edited by J. Fuller Maitland, W.B. Squire. Famous early 17th century collection of keyboard music, 300 works by Morley, Byrd, Bull, Gibbons, etc. Modern notation. Total of 938pp. 8³/8 x 11.
ECE 21068-5, 21069-3 Pa., Two vol. set $15.00

COMPLETE STRING QUARTETS, Wolfgang A. Mozart. Breitkopf and Härtel edition. All 23 string quartets plus alternate slow movement to K156. Study score. 277pp. 9³/8 x 12¼.
22372-8 Pa. $6.00

COMPLETE SONG CYCLES, Franz Schubert. Complete piano, vocal music of Die Schöne Müllerin, Die Winterreise, Schwanengesang. Also Drinker English singing translations. Breitkopf and Härtel edition. 217pp. 9³/8 x 12¼.
22649-2 Pa. $4.50

THE COMPLETE PRELUDES AND ETUDES FOR PIANOFORTE SOLO, Alexander Scriabin. All the preludes and etudes including many perfectly spun miniatures. Edited by K.N. Igumnov and Y.I. Mil'shteyn. 250pp. 9 x 12.
22919-X Pa. $5.00

TRISTAN UND ISOLDE, Richard Wagner. Full orchestral score with complete instrumentation. Do not confuse with piano reduction. Commentary by Felix Mottl, great Wagnerian conductor and scholar. Study score. 655pp. 8¹/8 x 11.
22915-7 Pa. $11.95

FAVORITE SONGS OF THE NINETIES, ed. Robert Fremont. Full reproduction, including covers, of 88 favorites: Ta-Ra-Ra-Boom-De-Aye, The Band Played On, Bird in a Gilded Cage, Under the Bamboo Tree, After the Ball, etc. 401pp. 9 x 12.
EBE 21536-9 Pa. $6.95

SOUSA'S GREAT MARCHES IN PIANO TRANSCRIPTION: ORIGINAL SHEET MUSIC OF 23 WORKS, John Philip Sousa. Selected by Lester S. Levy. Playing edition includes: The Stars and Stripes Forever, The Thunderer, The Gladiator, King Cotton, Washington Post, much more. 24 illustrations. 111pp. 9 x 12.
USO 23132-1 Pa. $3.50

CLASSIC PIANO RAGS, selected with an introduction by Rudi Blesh. Best ragtime music (1897-1922) by Scott Joplin, James Scott, Joseph F. Lamb, Tom Turpin, 9 others. Printed from best original sheet music, plus covers. 364pp. 9 x 12.
EBE 20469-3 Pa. $6.95

ANALYSIS OF CHINESE CHARACTERS, C.D. Wilder, J.H. Ingram. 1000 most important characters analyzed according to primitives, phonetics, historical development. Traditional method offers mnemonic aid to beginner, intermediate student of Chinese, Japanese. 365pp.
23045-7 Pa. $4.00

MODERN CHINESE: A BASIC COURSE, Faculty of Peking University. Self study, classroom course in modern Mandarin. Records contain phonetics, vocabulary, sentences, lessons. 249 page book contains all recorded text, translations, grammar, vocabulary, exercises. Best course on market. 3 12" 33¹/3 monaural records, book, album.
98832-5 Set $12.50

THE BEST DR. THORNDYKE DETECTIVE STORIES, R. Austin Freeman. The Case of Oscar Brodski, The Moabite Cipher, and 5 other favorites featuring the great scientific detective, plus his long-believed-lost first adventure — 31 New Inn — reprinted here for the first time. Edited by E.F. Bleiler. USO 20388-3 Pa. $3.00

BEST "THINKING MACHINE" DETECTIVE STORIES, Jacques Futrelle. The Problem of Cell 13 and 11 other stories about Prof. Augustus S.F.X. Van Dusen, including two "lost" stories. First reprinting of several. Edited by E.F. Bleiler. 241pp. 20537-1 Pa. $3.00

UNCLE SILAS, J. Sheridan LeFanu. Victorian Gothic mystery novel, considered by many best of period, even better than Collins or Dickens. Wonderful psychological terror. Introduction by Frederick Shroyer. 436pp. 21715-9 Pa. $4.00

BEST DR. POGGIOLI DETECTIVE STORIES, T.S. Stribling. 15 best stories from EQMM and The Saint offer new adventures in Mexico, Florida, Tennessee hills as Poggioli unravels mysteries and combats Count Jalacki. 217pp. 23227-1 Pa. $3.00

EIGHT DIME NOVELS, selected with an introduction by E.F. Bleiler. Adventures of Old King Brady, Frank James, Nick Carter, Deadwood Dick, Buffalo Bill, The Steam Man, Frank Merriwell, and Horatio Alger — 1877 to 1905. Important, entertaining popular literature in facsimile reprint, with original covers. 190pp. 9 x 12. 22975-0 Pa. $3.50

ALICE'S ADVENTURES UNDER GROUND, Lewis Carroll. Facsimile of ms. Carroll gave Alice Liddell in 1864. Different in many ways from final Alice. Handlettered, illustrated by Carroll. Introduction by Martin Gardner. 128pp. 21482-6 Pa. $1.50

ALICE IN WONDERLAND COLORING BOOK, Lewis Carroll. Pictures by John Tenniel. Large-size versions of the famous illustrations of Alice, Cheshire Cat, Mad Hatter and all the others, waiting for your crayons. Abridged text. 36 illustrations. 64pp. 8¼ x 11. 22853-3 Pa. $1.50

AVENTURES D'ALICE AU PAYS DES MERVEILLES, Lewis Carroll. Bué's translation of "Alice" into French, supervised by Carroll himself. Novel way to learn language. (No English text.) 42 Tenniel illustrations. 196pp. 22836-3 Pa. $2.50

MYTHS AND FOLK TALES OF IRELAND, Jeremiah Curtin. 11 stories that are Irish versions of European fairy tales and 9 stories from the Fenian cycle — 20 tales of legend and magic that comprise an essential work in the history of folklore. 256pp. 22430-9 Pa. $3.00

EAST O' THE SUN AND WEST O' THE MOON, George W. Dasent. Only full edition of favorite, wonderful Norwegian fairytales — Why the Sea is Salt, Boots and the Troll, etc. — with 77 illustrations by Kittelsen & Werenskiöld. 418pp. 22521-6 Pa. $4.00

PERRAULT'S FAIRY TALES, Charles Perrault and Gustave Doré. Original versions of Cinderella, Sleeping Beauty, Little Red Riding Hood, etc. in best translation, with 34 wonderful illustrations by Gustave Doré. 117pp. 8⅛ x 11. 22311-6 Pa. $2.50

MOTHER GOOSE'S MELODIES. Facsimile of fabulously rare Munroe and Francis "copyright 1833" Boston edition. Familiar and unusual rhymes, wonderful old woodcut illustrations. Edited by E.F. Bleiler. 128pp. 4½ x 6⅜. 22577-1 Pa. $1.50

MOTHER GOOSE IN HIEROGLYPHICS. Favorite nursery rhymes presented in rebus form for children. Fascinating 1849 edition reproduced in toto, with key. Introduction by E.F. Bleiler. About 400 woodcuts. 64pp. 6⅞ x 5¼. 20745-5 Pa. $1.00

PETER PIPER'S PRACTICAL PRINCIPLES OF PLAIN & PERFECT PRONUNCIATION. Alliterative jingles and tongue-twisters. Reproduction in full of 1830 first American edition. 25 spirited woodcuts. 32pp. 4½ x 6⅜. 22560-7 Pa. $1.00

MARMADUKE MULTIPLY'S MERRY METHOD OF MAKING MINOR MATHEMATICIANS. Fellow to Peter Piper, it teaches multiplication table by catchy rhymes and woodcuts. 1841 Munroe & Francis edition. Edited by E.F. Bleiler. 103pp. 4⅝ x 6.
22773-1 Pa. $1.25
20171-6 Clothbd. $3.00

THE NIGHT BEFORE CHRISTMAS, Clement Moore. Full text, and woodcuts from original 1848 book. Also critical, historical material. 19 illustrations. 40pp. 4⅝ x 6. 22797-9 Pa. $1.25

THE KING OF THE GOLDEN RIVER, John Ruskin. Victorian children's classic of three brothers, their attempts to reach the Golden River, what becomes of them. Facsimile of original 1889 edition. 22 illustrations. 56pp. 4⅝ x 6⅜.
20066-3 Pa. $1.50

DREAMS OF THE RAREBIT FIEND, Winsor McCay. Pioneer cartoon strip, unexcelled for beauty, imagination, in 60 full sequences. Incredible technical virtuosity, wonderful visual wit. Historical introduction. 62pp. 8⅜ x 11¼. 21347-1 Pa. $2.50

THE KATZENJAMMER KIDS, Rudolf Dirks. In full color, 14 strips from 1906-7; full of imagination, characteristic humor. Classic of great historical importance. Introduction by August Derleth. 32pp. 9¼ x 12¼. 23005-8 Pa. $2.00

LITTLE ORPHAN ANNIE AND LITTLE ORPHAN ANNIE IN COSMIC CITY, Harold Gray. Two great sequences from the early strips: our curly-haired heroine defends the Warbucks' financial empire and, then, takes on meanie Phineas P. Pinchpenny. Leapin' lizards! 178pp. 6⅛ x 8⅜. 23107-0 Pa. $2.00

THE BEST OF GLUYAS WILLIAMS. 100 drawings by one of America's finest cartoonists: The Day a Cake of Ivory Soap Sank at Proctor & Gamble's, At the Life Insurance Agents' Banquet, and many other gems from the 20's and 30's. 118pp. 8⅜ x 11¼. 22737-5 Pa. $2.50

DRIED FLOWERS, Sarah Whitlock and Martha Rankin. Concise, clear, practical guide to dehydration, glycerinizing, pressing plant material, and more. Covers use of silica gel. 12 drawings. Originally titled "New Techniques with Dried Flowers." 32pp. 21802-3 Pa. $1.00

ABC OF POULTRY RAISING, J.H. Florea. Poultry expert, editor tells how to raise chickens on home or small business basis. Breeds, feeding, housing, laying, etc. Very concrete, practical. 50 illustrations. 256pp. 23201-8 Pa. $3.00

HOW INDIANS USE WILD PLANTS FOR FOOD, MEDICINE & CRAFTS, Frances Densmore. Smithsonian, Bureau of American Ethnology report presents wealth of material on nearly 200 plants used by Chippewas of Minnesota and Wisconsin. 33 plates plus 122pp. of text. 6⅛ x 9¼. 23019-8 Pa. $2.50

THE HERBAL OR GENERAL HISTORY OF PLANTS, John Gerard. The 1633 edition revised and enlarged by Thomas Johnson. Containing almost 2850 plant descriptions and 2705 superb illustrations, Gerard's Herbal is a monumental work, the book all modern English herbals are derived from, and the one herbal every serious enthusiast should have in its entirety. Original editions are worth perhaps $750. 1678pp. 8½ x 12¼. 23147-X Clothbd. $50.00

A MODERN HERBAL, Margaret Grieve. Much the fullest, most exact, most useful compilation of herbal material. Gigantic alphabetical encyclopedia, from aconite to zedoary, gives botanical information, medical properties, folklore, economic uses, and much else. Indispensable to serious reader. 161 illustrations. 888pp. 6½ x 9¼. USO 22798-7, 22799-5 Pa., Two vol. set $10.00

HOW TO KNOW THE FERNS, Frances T. Parsons. Delightful classic. Identification, fern lore, for Eastern and Central U.S.A. Has introduced thousands to interesting life form. 99 illustrations. 215pp. 20740-4 Pa. $2.75

THE MUSHROOM HANDBOOK, Louis C.C. Krieger. Still the best popular handbook. Full descriptions of 259 species, extremely thorough text, habitats, luminescence, poisons, folklore, etc. 32 color plates; 126 other illustrations. 560pp. 21861-9 Pa. $4.50

HOW TO KNOW THE WILD FRUITS, Maude G. Peterson. Classic guide covers nearly 200 trees, shrubs, smaller plants of the U.S. arranged by color of fruit and then by family. Full text provides names, descriptions, edibility, uses. 80 illustrations. 400pp. 22943-2 Pa. $4.00

COMMON WEEDS OF THE UNITED STATES, U.S. Department of Agriculture. Covers 220 important weeds with illustration, maps, botanical information, plant lore for each. Over 225 illustrations. 463pp. 6⅛ x 9¼. 20504-5 Pa. $4.50

HOW TO KNOW THE WILD FLOWERS, Mrs. William S. Dana. Still best popular book for East and Central USA. Over 500 plants easily identified, with plant lore; arranged according to color and flowering time. 174 plates. 459pp. 20332-8 Pa. $3.50

THE STYLE OF PALESTRINA AND THE DISSONANCE, Knud Jeppesen. Standard analysis of rhythm, line, harmony, accented and unaccented dissonances. Also pre-Palestrina dissonances. 306pp. 22386-8 Pa. $4.50

DOVER OPERA GUIDE AND LIBRETTO SERIES prepared by Ellen H. Bleiler. Each volume contains everything needed for background, complete enjoyment: complete libretto, new English translation with all repeats, biography of composer and librettist, early performance history, musical lore, much else. All volumes lavishly illustrated with performance photos, portraits, similar material. Do not confuse with skimpy performance booklets.

CARMEN, Georges Bizet. 66 illustrations. 222pp. 22111-3 Pa. $3.00
DON GIOVANNI, Wolfgang A. Mozart. 92 illustrations. 209pp. 21134-7 Pa. $2.50
LA BOHÈME, Giacomo Puccini. 73 illustrations. 124pp. USO 20404-9 Pa. $1.75
ÄIDA, Giuseppe Verdi. 76 illustrations. 181pp. 20405-7 Pa. $2.25
LUCIA DI LAMMERMOOR, Gaetano Donizetti. 44 illustrations. 186pp. 22110-5 Pa. $2.00

ANTONIO STRADIVARI: HIS LIFE AND WORK, W. H. Hill, et al. Great work of musicology. Construction methods, woods, varnishes, known instruments, types of instruments, life, special features. Introduction by Sydney Beck. 98 illustrations, plus 4 color plates. 315pp. 20425-1 Pa. $4.00

MUSIC FOR THE PIANO, James Friskin, Irwin Freundlich. Both famous, little-known compositions; 1500 to 1950's. Listing, description, classification, technical aspects for student, teacher, performer. Indispensable for enlarging repertory. 448pp. 22918-1 Pa. $4.00

PIANOS AND THEIR MAKERS, Alfred Dolge. Leading inventor offers full history of piano technology, earliest models to 1910. Types, makers, components, mechanisms, musical aspects. Very strong on offtrail models, inventions; also player pianos. 300 illustrations. 581pp. 22856-8 Pa. $5.00

KEYBOARD MUSIC, J.S. Bach. Bach-Gesellschaft edition. For harpsichord, piano, other keyboard instruments. English Suites, French Suites, Six Partitas, Goldberg Variations, Two-Part Inventions, Three-Part Sinfonias. 312pp. 8⅛ x 11. 22360-4 Pa. $5.00

COMPLETE STRING QUARTETS, Ludwig van Beethoven. Breitkopf and Härtel edition. 6 quartets of Opus 18; 3 quartets of Opus 59; Opera 74, 95, 127, 130, 131, 132, 135 and Grosse Fuge. Study score. 434pp. 9⅜ x 12¼. 22361-2 Pa. $7.95

COMPLETE PIANO SONATAS AND VARIATIONS FOR SOLO PIANO, Johannes Brahms. All sonatas, five variations on themes from Schumann, Paganini, Handel, etc. Vienna Gesellschaft der Musikfreunde edition. 178pp. 9 x 12. 22650-6 Pa. $4.50

PIANO MUSIC 1888-1905, Claude Debussy. Deux Arabesques, Suite Bergamesque, Masques, 1st series of Images, etc. 9 others, in corrected editions. 175pp. 9⅜ x 12¼. 22771-5 Pa. $4.00

INCIDENTS OF TRAVEL IN YUCATAN, John L. Stephens. Classic (1843) exploration of jungles of Yucatan, looking for evidences of Maya civilization. Travel adventures, Mexican and Indian culture, etc. Total of 669pp.
20926-1, 20927-X Pa., Two vol. set $6.00

LIVING MY LIFE, Emma Goldman. Candid, no holds barred account by foremost American anarchist: her own life, anarchist movement, famous contemporaries, ideas and their impact. Struggles and confrontations in America, plus deportation to U.S.S.R. Shocking inside account of persecution of anarchists under Lenin. 13 plates. Total of 944pp. 22543-7, 22544-5 Pa., Two vol. set $9.00

AMERICAN INDIANS, George Catlin. Classic account of life among Plains Indians: ceremonies, hunt, warfare, etc. Dover edition reproduces for first time all original paintings. 312 plates. 572pp. of text. 6⅛ x 9¼.
22118-0, 22119-9 Pa., Two vol. set $8.00
22140-7, 22144-X Clothbd., Two vol. set $16.00

THE INDIANS' BOOK, Natalie Curtis. Lore, music, narratives, drawings by Indians, collected from cultures of U.S.A. 149 songs in full notation. 45 illustrations. 583pp. 6⅝ x 9⅜. 21939-9 Pa. $6.95

INDIAN BLANKETS AND THEIR MAKERS, George Wharton James. History, old style wool blankets, changes brought about by traders, symbolism of design and color, a Navajo weaver at work, outline blanket, Kachina blankets, more. Emphasis on Navajo. 130 illustrations, 32 in color. 230pp. 6⅛ x 9¼. 22996-3 Pa. $5.00
23068-6 Clothbd. $10.00

AN INTRODUCTION TO THE STUDY OF THE MAYA HIEROGLYPHS, Sylvanus Griswold Morley. Classic study by one of the truly great figures in hieroglyph research. Still the best introduction for the student for reading Maya hieroglyphs. New introduction by J. Eric S. Thompson. 117 illustrations. 284pp. 23108-9 Pa. $4.00

THE ANALECTS OF CONFUCIUS, THE GREAT LEARNING, DOCTRINE OF THE MEAN, Confucius. Edited by James Legge. Full Chinese text, standard English translation on same page, Chinese commentators, editor's annotations; dictionary of characters at rear, plus grammatical comment. Finest edition anywhere of one of world's greatest thinkers. 503pp. 22746-4 Pa. $5.00

THE I CHING (THE BOOK OF CHANGES), translated by James Legge. Complete translation of basic text plus appendices by Confucius, and Chinese commentary of most penetrating divination manual ever prepared. Indispensable to study of early Oriental civilizations, to modern inquiring reader. 448pp.
21062-6 Pa. $3.50

THE EGYPTIAN BOOK OF THE DEAD, E.A. Wallis Budge. Complete reproduction of Ani's papyrus, finest ever found. Full hieroglyphic text, interlinear transliteration, word for word translation, smooth translation. Basic work, for Egyptology, for modern study of psychic matters. Total of 533pp. 6½ x 9¼.
EBE 21866-X Pa. $4.95